Global Environmental Studies

Editor-in-Chief:
Ken-ichi Abe

Series Editors:
Daniel Niles
Hein Mallee

More information about this series at http://www.springer.com/series/10124

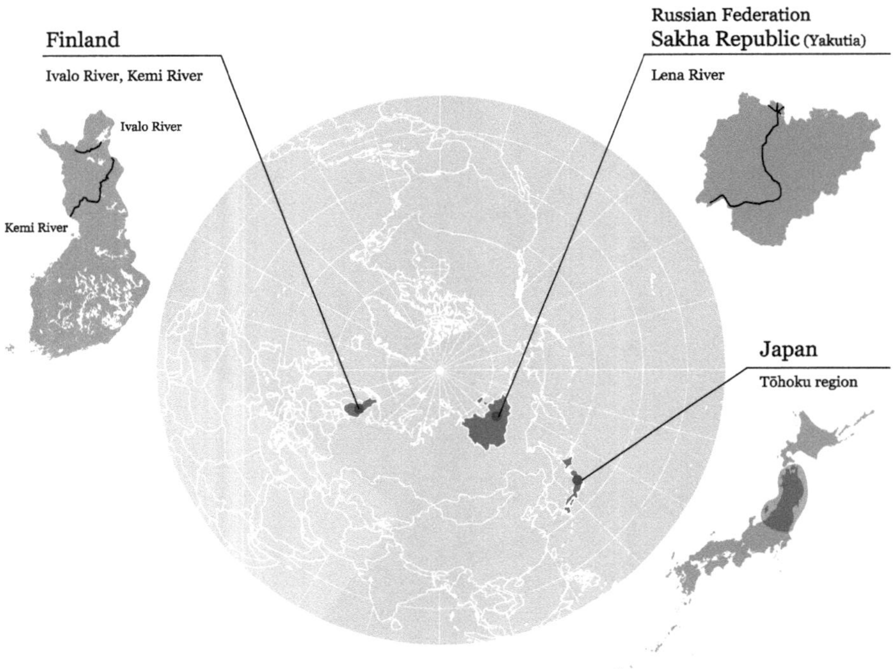
Finland
Ivalo River, Kemi River
Ivalo River
Kemi River
Russian Federation
Sakha Republic (Yakutia)
Lena River
Japan
Tōhoku region

Tetsuya Hiyama • Hiroki Takakura
Editors

Global Warming and Human - Nature Dimension in Northern Eurasia

Springer

Editors
Tetsuya Hiyama
Institute for Space-Earth Environmental Research
Nagoya University
Nagoya, Aichi, Japan

Hiroki Takakura
Center for Northeast Asian Studies
Tohoku University
Sendai, Miyagi, Japan

ISSN 2192-6336 ISSN 2192-6344 (electronic)
Global Environmental Studies
ISBN 978-981-10-4647-6 ISBN 978-981-10-4648-3 (eBook)
DOI 10.1007/978-981-10-4648-3

Library of Congress Control Number: 2017943114

Printed on acid-free paper

This Springer imprint is published by Springer Nature
The registered company is Springer Nature Singapore Pte Ltd.
The registered company address is: 152 Beach Road, #21-01/04 Gateway East, Singapore 189721, Singapore

Preface

This book is part of the series *Global Environmental Studies*, from Springer, supported by the Research Institute for Humanity and Nature (RIHN), a leading institute promoting interdisciplinary and transdisciplinary global socio-ecological research in Japan.

This book aims to describe the current status of environmental change caused by global warming in Northern Eurasia, with a special focus on Eastern Siberia. It concentrates on spring river flooding, ice-jam movement, and the monitoring of both conditions using field observations and remote sensing. The issues facing reindeer herders in Siberia and conspicuous environmental changes such as waterlogging, rising temperatures, and vegetation change are also discussed. In addition, adaptation strategies implemented by the government at various levels are also discussed.

The primary topics of this book are (i) an introduction to global warming and the human–nature dynamic in Siberia, with special emphasis on humidification of the region in the mid-2000s; (ii) a description of social adaptation to changes in the terrestrial ecosystem, with an emphasis on water environments; and (iii) a discussion of adaptation strategies based on an assessment of vulnerability to environmental change in Northern Eurasia. The latter topic is presented as a local phenomenon, influenced by climate science politics among states and intergovernmental organizations such as the Intergovernmental Panel on Climate Change (IPCC). The book covers scientific findings of studies in hydrology, ecology, anthropology, politics, and civil engineering and provides a multifaceted approach including information about ongoing processes as well as insightful theoretical considerations. We believe that the scientific analysis in this book will be of interest not only to environmental researchers and policy makers but also to local people who are affected either as right-holders or as stakeholders.

A discussion on the background of this book follows. The main contents (Chaps. 1, 2, 3, 4 and 9, 10, 11) are based on the first international conference on "Global Warming and the Human–Nature Dimension in Siberia: Social Adaptation to the Changes of the Terrestrial Ecosystem, with an Emphasis on Water Environments," which was held in Kyoto, Japan, on March 7–9, 2012. It examined human–nature interactions affected by environmental changes in Siberia. The foci of the conference included (i) contemporary and future variations in the water and carbon cycles, (ii) the results of long-term field observations of the effects of carbon and hydrologic variability and the key driving forces behind these effects, and (iii) the distinct social economies of multiethnic Siberian societies and their potential for adaptation to predicted changes in climate and terrestrial ecosystems. These arguments are laid out in the aforementioned chapters.

Chapters 8, 12, and 13 are based on the RIHN's eighth international symposium, entitled "Risk Societies, Edge Environments: Ecosystems and Livelihoods in the Balance," which was held in Kyoto, Japan, on October 23–25, 2013. This symposium examined social and ecological risk in several "edge" environments, i.e., boundary zones that typically exhibit high rates of biodiversity and many ecological niches, but that are also particularly susceptible to disturbance. The editors of this book organized the session entitled "Global Warming Risk in the Far North" at this symposium. It addressed the social and ecological risks presented by global warming in the circumpolar North, including Russia.

Over millennia, Northern indigenous peoples have adapted to a cold climate, but their cultural traditions and techniques are being disrupted by natural disasters related to climate change. Modern global warming has apparently increased the magnitude and frequency of flooding and land erosion. Thus, new adaptation strategies are needed in relation to these risks and should be explored. The case studies in this book show how indigenous peoples, local governments, and civil organizations evaluate contemporary environmental change and mitigate its risks.

It should be noted that a research project with the same title as the first symposium "Global Warming and the Human–Nature Dimension in Siberia: Social Adaptation to the Changes of the Terrestrial Ecosystem, with an Emphasis on Water Environments" took place between 2009 and 2014 at RIHN. One of the editors, Tetsuya Hiyama, led this project. Most of the members of the project had not previously experienced a multidisciplinary approach including natural and social sciences. We remember clearly the misunderstandings and anxieties among the team members during the first 1–2 years. After many discussions at research meetings and/or the shared experience of fieldwork, the tensions were eventually mitigated, and we came to a mutual understanding. On a personal note, most members of the project experienced the Great East Japan Earthquake on March 11, 2011, together. We had a research meeting at Tohoku University, Sendai, near one of epicenters of the quake on that day. The shared experience of the disaster and evacuation contributed to the sense of solidarity and commitment among the project members.

This book successfully uses a crosscutting approach to the issue of Northern Eurasian climate change, based on its background as a collaborative project between scholars in the natural sciences and human social sciences. The international collaboration included Russian, Finnish, American, and Japanese contributors from both natural and social sciences who tried to explore shared concerns. The editors believe that we have accomplished a well-balanced integration of multiple disciplines for exploring human–environment interactions affected by climate change.

The in-depth description of socio-ecological processes will be satisfying to readers interested in climate change science, Arctic studies, and Russian studies.

Nagoya, Japan Tetsuya Hiyama
Sendai, Japan Hiroki Takakura

Contents

Global Environmental Studies

The Global Environmental Studies series introduces the research undertaken at, or in association with, the Research Institute for Humanity and Nature (RIHN). Located in Kyoto, Japan, RIHN is a national institute conducting fixed-term, multi-disciplinary, international research projects on pressing areas of environmental concern.

RIHN seeks to transcend the common divisions between the humanities and the social and natural sciences, and to develop synthetic and transformative descriptions of humanity in the midst of a dynamic, changeable nature. The works published in the series will reflect the full breadth of RIHN scholarship in this transdisciplinary field of global environmental studies.

Chapter 1
Water, Water Everywhere: Perceptions of Chaotic Water Regimes in Northeastern Siberia, Russia

Susan A. Crate

Abstract This paper explores applied anthropological research examining perceptions, understandings and responses to increasing water on the land, one of the major effects of global climate change for native Viliui Sakha agropastoralist communities of northeastern Siberia, Russia. The paper draws on fieldwork investigating perceptions, understandings and responses to the local effects of global climate change for native Viliui Sakha agropastoralist communities of northeastern Siberia, Russia. For Viliui Sakha, global climate change translates locally into a highly altered climate system and water regime. 2008 fieldwork shows Inhabitants observing warmer winters, increased snowfall, excessive precipitation, changed seasonality, and the transformation of their ancestral landscape due to increased water on the land and degrading permafrost. One urgent change is how the increased water on the land is turning hayfields into lakes, inundating households and ruining transportation networks. The increasing water on the land interferes with subsistence and threatens to undermine settlement. Beyond these physical changes, what does the increased water on the land mean to Viliui Sakha? Inhabitants expressed not only concern about their future but also common fear that they would 'go under water.' Water has visceral meaning to Sakha, based on their historically-based belief system, their adaptation to their environment, and knowledge system. In response, 2009 field research looked in more depth at communities' perceptions of water, and worked to bring those perceptions and beliefs into our 2010 knowledge exchange exercise. This paper will present our initial findings and make suggestions on how these findings can be understood more broadly for other peoples unprecedentedly affected by water crises in the face of global climate change.

Keywords Climate change • Anthropology • Local perceptions • Water • Viliui Sakha • Siberia • Russia

---

S.A. Crate (✉)
George Mason University, Fairfax, VA, USA
e-mail: scrate1@gmu.edu

T. Hiyama, H. Takakura (eds.), *Global Warming and Human - Nature Dimension in Northern Eurasia*, Global Environmental Studies,
DOI 10.1007/978-981-10-4648-3_1

1.1 Introduction

> The climate is definitely different from before ... For people who live with a short summer when there needs to be the right weather to accomplish all for the winter, and there are cool rainy times so that the hay does not dry and has to sit and sit and the quality is bad because of that ... it is the right time for haying but the conditions are all wrong. – Sakha elder

At the end of a 2003–2005 research project[1] focusing on understanding local definitions of sustainability in Viliui Sakha villages of northeastern Siberia, Russia (Crate 2006a), our research team administered a survey to gauge the extent that what inhabitants expressed concern about in focus groups and semi-structured interviews was shared by the overall population. The final question asked if the respondent had any other concerns that the survey did not cover. Ninety percent of the responses to that question had something to do with changes in local weather, climate and seasonal timing. In response, our team used the final 2 weeks of the summer's field research to interview elders, those having the longest life experience observing change, about their observations. In the context of those interviews, remarks like the one above by a Sakha elder were common. In addition to providing lifetime observations of these changes, the elders also sparked our team's curiosity about how a people are affected culturally by climate change, in ways beyond the physical realm. Ten of the 33 elders interviewed explained that these unprecedented changes were a result of the Bull of Winter no long arriving (Crate 2008).

The Bull of Winter is a mythological being who Sakha believe arrives when the deep Siberian winter, characterized by conditions too cold and dry for snow and relatively windless, in late December. Those ten elders stated that the Bull of Winter was not arriving in the last 10 years, with overall temperatures softening. This led me, from my training as an anthropologist and therefore keen to pursue issues of culture and change, to want to understand the cultural implications of climate change for Viliui Sakha. At the time, I had been working with these people since 1991 on various issues of change and adaptation (Crate 2006b). I wanted to know how they would continue to inhabit their homelands if climate change altered their environment to the extent that they could no longer breed horses and cows, their age-old subsistence strategy. Perhaps more significant in my mind was pondering the meaning and power of place for Viliui Sakha and the implications of that place changing to the extent that it no longer supported familiar ecosystem conditions and cosmological meanings. While pondering these issues, I confronted sentiments like

[1] I first thank all Viliui region inhabitants, project collaborators, research assistants, and in-country specialists involved in the research on which this article is founded. I also acknowledge the National Science Foundation (NSF) and program officers Anna Kertulla de Echave, Office of Polar Programs, Arctic Social Sciences Division, and Neil Swanberg, Office of Polar Programs, Arctic Sciences Division, for funding support. This article is mainly based on research from the NSF Office of Polar Programs, Arctic Social Science Program Grant 0710935 "Assessing Knowledge, Resilience & Adaptation and Policy Needs in Northern Russian Villages Experiencing Unprecedented Climate Change," and NSF Office of Polar Programs, Arctic Science Program Grant 0902146 "Understanding Climate-Driven Phenological Change: Observations, Adaptations and Cultural Implications in Northeastern Siberia and Labrador/Nunatsiavut (PHENARC)."

the following that made my curiosity that much stronger: "The Eveny people are highly adaptive. Sometimes they joke and say, this is our home. If the climate gets too hot, we'll just stay and herd camels" (anonymous). There is no doubt that Viliui Sakha, like their neighbors the Eveny, are resilient, adaptive cultures, but the question remained as to how much environmental change they could adapt to.

Anthropologists have long argued for the importance to local peoples of the subtler and non-physical characteristics of local ecosystems. For example, Keith Basso argues that social life is everywhere accomplished through an exchange of symbolic forms, that human existence is irrevocably situated in time and space, and that wisdom "sits in places" (Basso 1996: 53). In agreement with Basso's contention, it follows that climate change can displace not only a people's wisdom but also the very human-environment interactions that are a culture's core (Steward 1955; Netting 1968, 1993).

1.2 Investigating Change

To pursue this investigation in-depth, I submitted a specific project investigating perceptions, understandings and responses to the local effects of global climate change, and the National Science Foundation (NSF) funded it (Fig. 1.1).

Our research team first organized focus groups to gain a solid understanding of what changes local inhabitants were observing, how they were understanding those changes, and how those changes were affecting their lives. We worked in four Viliui Sakha villages in the Suntar region, namely Elgeeii, Kutana, Khoro and Tolon. In each, we hired a village assistant to identify six women and six men, with an even breakdown of youth, middle age, and elderly participants, for each focus group. In these group interviews, we first asked participants to complete a form that asked them to record their observations of change, what they thought the source of that change was, how it was affecting them, and what they thought the future would be if the change continued. Through analysis of focus group results, we discerned that the majority of participants were in agreement about the following nine main changes: (1) winters are warm; (2) the land is waterlogged; (3) too much rain; (4) summers are cold; (5) more floods; (6) seasons arrive late; (7) too much snow; (8) temperatures change more suddenly; and, (9) fewer birds and animals. Although inhabitants were quick to identify that warmer winters are a welcome change because it means less wood is required to heat their homes, and animals could be outside more and go to water instead of having to haul it for them, the graver implications of these changes tended to dominate our discussions. In the end, we determined that each of these changes has largely negative effects on Viliui Sakha's horse and cattle subsistence, in addition to creating problems with transportation, housing and supplemental subsistence like gardening, hunting, and fishing.

Most of the changes listed represent the effects of a highly altered climate system and water regime. Like all humans in the context of their local environment, Sakha's ancestors adapted to an annual water regime that had an overall sufficient supply to

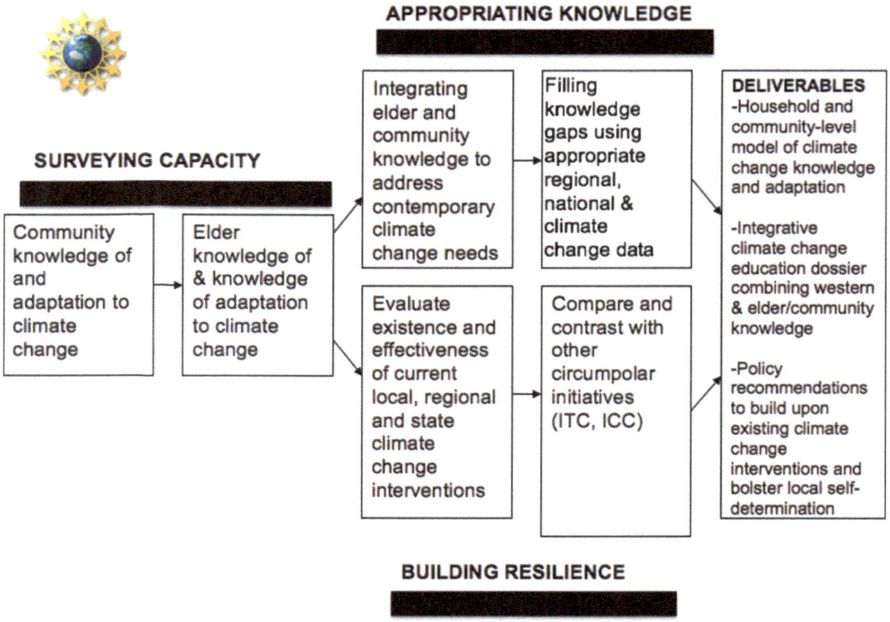

Fig. 1.1 Assessing knowledge, resilience & adaptation and policy needs in Viliui Sakha villages experiencing unprecedented climate change

support ecosystem functioning but that also fluctuated from year to year in a relatively predictable range of wet and dry. However, with the advent of contemporary climate change, this range is increasingly disrupted, due to a combination of changing seasonal patterns, altered precipitation regimes and an overall "softening" of the extreme annual temperature range (Crate 2011a, b; Roshydromet 2008; Fedorov and Konstantinov 2008, 2009). By increasing labor and paying higher costs, inhabitants are adapting to these changes. However, the number of households continuing to keep cows has dropped precipitously in the last 5 years, suggesting that fewer are meeting the increasing challenge.

As I worked with inhabitants in subsequent years, they talked frequently about warmer winters, increased snowfall, excessive precipitation, changed seasonality, and the transformation of their ancestral landscape due to increasing water on the land and degrading permafrost. One urgent change is how the increased water on the land is turning hayfields into lakes, inundating households, and ruining transportation networks. The increasing water interferes with subsistence and threatens to undermine settlements (Fig. 1.2).

Interviews in 2012 confirmed previously reported seasonal changes including: spring either arriving late or on time but remaining cold or not gradually warming up as in the past; less rain in the spring than before; a cool summer with too much rain during hay season and cold nights; an elongated autumn characterized by a freeze – thaw pattern; an unusual winter season, with more snow than previously

Fig. 1.2 "Like this, 10 years of water I do not remember – before we hayed all the fields – now we have to go here and there to hay since all our hay lands are under water." (middle-aged Viliui Sakha resident)

and *chiskhaan* or freezing winds. These seasonal changes challenge Viliui Sakha in their cow-keeping and other subsistence practices. Because of a lack of space here to go into detail about all the effects, for the sake of illustration, I next highlight a few for each season.

Inhabitants emphasized that the elongated autumn is better because cows can go to pasture later which prolongs winter hay stores. However, it also requires more hay stores since pasture is often not available into winter and the warm temperatures delay the timing of the annual slaughter, performed after temperatures have fallen sufficiently to remain below freezing. Horses are negatively affected because the freeze – thaw pattern keeps them from accessing their fodder, which typically remains semi-green beneath a thick, insulating blanket of snow. Instead, they encounter a hard sheet of ice and, unable to find fodder, many starve if not given supplemental food. The winter is unusual due to snowfall and freezing wind in an ecosystem in which the historical pattern is no snowfall during that 3-month season and no wind, the period Sakha characterize as when the Bull of Winter has arrived. This new regime requires much more human labor, since cattle stay in barns for more of the winter due to uncertain conditions of their paths to water. Spring is late or arrives on time but remains cold instead of a gradual warming as before, and there is less rain. These extreme spring conditions affect early fundamental hay growth because of the cold and lack of rain. Recent summers have been unseasonably cool and rainy and the nights are especially cold, slowing plant growth, meaning that harvest time now comes before crops have reached full maturity. Last, summer used to be a relatively dry season, but now there is too much rain, which impedes hay cutting and ruins its quality.

Beyond these physical changes, what does the increased water on the land mean to Viliui Sakha? Inhabitants expressed not only concern about their future but also a common fear that they would "go under water". One middle-aged Viliui Sakha inhabitant said, "I am very scared that we are going under water—looking down from a plane you can see that the land has patches of water across it—water is coming up from below—it looks like the land is sinking down." Water has visceral meaning to the Sakha, based on their historically-based belief system, their adaptation to the environment, and knowledge system. Scholars specializing in the meaning of water to various peoples are outspoken about how important it is to understand those meanings to grasp perceptions. A people's interactions with water, and how experiences, images and metaphors arise from those interactions and flow into their interpretations of themselves and others (Strang 2004).

In response to both Viliui Sakha's expressions of fear and the experts' call to investigate the meanings of water, our team continued field research, taking a more in-depth look at communities' perceptions of water. Based on Sakha's ancient belief system, water, like all of the natural world, is sentient or spirit-filled. For example, when interacting with water, for example, when crossing a river or fishing in a lake, according to their ancient belief, they are to speak specific words and offer gifts to appease the spirit of the water. As an anthropologist, I knew that it was important to grasp the "emic" or local understanding, similar to Arlene Rosen's argument in her 2007 book, *Civilizing Climate*:

> If rainfall is a divine gift, then solving the problems related to drought must involve dealings with the supernatural in the form of pleasing the deity responsible. Failure to adjust to environmental stress is as much a social and cosmological problem as an environmental one (Rosen 2007).

In addition, it was clear that inhabitants were attributing such changes to many other sources besides climate change. Granted, although none of the changes were occurring *solely* due to climate change, it was certainly a driving force for most. When asked what they thought was causing the changes, their main explanations, in order of preference, were: (1) the Viliui hydroelectric station reservoir; (2) nature itself, i.e., wet years/ dry year cycles; (3) too much *technika* (explained later); and, (4) global climate change. Regarding the first explanation, inhabitants explained how the huge water surface of the reservoir generates steam that forms clouds that enter their area, keeping their climate artificially cool in summer and warm in winter, and increasing rain and snow year round. A more recent opening of GES's third generator, which further expanded reservoir volume and surface area, was responsible for an increase in these effects over recent years. However, hydrological and meteorological research shows how the reservoir affects only a microclimate area directly adjacent to it (Shadrin 1984; Nogovitzin 1985).

The second explanation purports that the observed changes were because of a pre-existing wet and dry year "natural cycle," a normal condition of the ecosystem to which Viliui Sakha ancestors adapted. They did so by practicing a form of water management called *nulustur,* which entailed either draining or flooding productive areas to maintain proper growing conditions within the natural wet and dry cycles (Ermolaev 1991). Those who subscribed to this explanation held that there was nothing humans could do or, for that matter, should do. Some even said that the government's draining of the lakes is wrong because when the dry cycle returns, the lands will be rendered much too dry and will never recover. Several were concerned about angering the spirit of the water by excessive draining. When asked what to do to adapt to the excessive water, most said to rely on Sakha adaptations. Specifically they mentioned *muus oto* or cutting hay after lake and pond ice freezes, *wolba* or cutting hay in the water, feeding herds with other fodder including ground-up shrubs and trees, and moving households temporarily to the higher lands of headwaters to cut hay and pasture their herds.

The third explanation, "too much *technika*," refers to the increasing use of mechanization both locally and globally, as witnessed on television and in the news and in the activity of the night sky. Most residents who attributed contemporary change to this explanation were elders who had certainly seen great change in their lifetimes, beginning with a world powered by human and animal labor to the current cyber-reality. Finally, only a handful of the participants explained the changes as a result of contemporary climate change. These tended to be either teachers or elders who read extensively.

To see the extent to which our findings in the focus groups and interviews were shared in the greater communities, we administered a survey in summer 2009. The most relevant results for this chapter's purpose are that a majority agreed with the nine main changes and with the four explanations of causes of change. These explanations, with climate change being the least mentioned one, showed the extent to which perceptions are historically constructed, in Viliui Sakha's case, very much based on Soviet-period industrial development that severely altered Viliui Sakha's natural environment (Crate 2002). My further investigation into these explanations revealed how these people clarify the political ecology with "water in mind" (Crate 2011a). More pertinent to the immediate situation, I collaborated with Alexander Fedorov of the Melnikov Permafrost Institute of Yakutsk, whose extensive research data from the central Sakha Republic showed that permafrost degradation and many of the other changes Viliui Sakha were observing were partially or completely attributable to global climate change. Because my research had shown how inhabitants had very locale- and culture-specific knowledge of these changes, Alexander Fedorov and I decided that the next step was to explore the extent to which these two knowledge bases could complement each other and enhance understanding for both local residents and regional scientists. We envisioned a two-part process: (1) To develop models and frameworks to understand a people's mythological, cosmological, historical, geopolitical, and contemporary perceptions of and responses to climate change; (2) to provide sources of information that complement local understanding and facilitate adaptive resilience, using a research approach that is community-based, collaborative, and locale- and culture-specific.

Fig. 1.3 2010 Knowledge exchange team preparing to leave Yakutsk to conduct knowledge exchanges in eight Viliui Sakha settlements: Viliuisk, Verkhny-viliuisk, Nerba, Suntar, Elgeeii, Kutana, Khoro and Tolon

Therefore, we decided to conduct what we called "knowledge exchanges" in summer 2010 (Crate and Fedorov 2013a). We spent winter and spring 2010 preparing, and then departed in late June of that year to do knowledge exchanges in eight Viliui Sakha settlements (Fig. 1.3). We structured these exchanges to provide a maximum opportunity for the audience to share their local observations, first inviting audience members to do so. For illustrative purposes, I provide two excerpts of these testimonies here. The first is the testimony of a life-long resident of Khoro Village known for her vegetable gardening:

> As a gardener, I know that animals and birds changing; some are more and some are less. There are some new birds, the *druzhdi* (thrush) and *grach* (rook), and they make nests and have lived here for the last 5 or 6 years. They are moving in here and move from one village to the next each year. There were two nests outside my house and this year, it seems like 100 appeared. They eat the insects but also the tomatoes—the very ripest ones—and suck the juice out. If you keep the tomatoes in the greenhouse they cannot reach them but on the open ground they can. Never before have birds eaten our vegetables!

The second illustrative testimony is from a horse-watcher who traveled 20 km by horse to attend the knowledge exchange:

> I am a cowboy and hunter. The land has changed—there are now ravines and gullies where there never were before. The land has caved in—around the streams where the flat hay lands are, the same thing is happening; the ravines have formed and all the water goes away very quickly down them, and lands are now dry that were never dry before. The ducks come very late. I am speaking mostly about changes in the land's form; there are now hills or mounds

> where there never were before, and small holes or hollows where there never were before. The wetlands are now dry and the ravines are more and more—land that never had water now does. Erosion of our lands is occurring very quickly, over just the last few years. Horses have a very hard time in the warm winter; we thought that this would be good for them but now understand that it is bad. Overall, we have come to a time when we cannot forecast the weather; before, things came at the right times, but now it is not so.

In all eight settlements, these stories varied, but all are very illustrative of the diversity of peoples' perceptions and experiences and the different ways the local environment is being affected by this global phenomenon.

Following the audience testimonies, we presented our findings in collaboration with local communities to define the major changes, their effects and causes. When we shared these results, audience members were quite moved to understand how similar our research respondents' experiences were to theirs. This tended to prompt more audience sharing and discussion. Then, Alexander Fedorov presented his regional research on permafrost degradation, first explaining how global climate change is affecting the earth system overall, the specific ways in which Russia is affected, and the particulars within the Sakha Republic. He shared pictures of how the land was falling and rising in places because of the changing permafrost layer. He also showed ways to protect the permafrost by building new houses and retrofitting older ones. His presentation stimulated even more discussion and we entertained many questions.

1.3 Conclusion

In follow-up evaluations with knowledge exchange participants, we discovered an overwhelming interest in bringing the knowledge exchange process to all inhabitants of the Viliui regions. In response, we completed a handbook designed to lead readers through a very similar process as the knowledge exchanges, and incorporates local knowledge, our research findings, and Fedorov's explanations of global climate change and permafrost degradation (Crate and Fedorov 2013b). With our research funding we were able to publish 3000 copies and, with coordination by the Sakha Monistry of Ecology, distribute the handbooks throughout the Viliui regions. We hope that we can share this knowledge exchange model with other communities facing the same issues, toward increasing understanding on the local level and bolstering adaptive responses.

References

Basso K (1996) Wisdom sits in places: notes on a western Apache landscape. In: Basso K, Feld S (eds) Senses of place. School of American Research Press, Santa Fe, pp 53–90

Crate SA (2002) Co-option in Siberia: the case of diamonds the Vilyuy Sakha. Polar Geogr 26(4):289–307

Crate SA (2006a) Investigating local definitions of sustainability in the Arctic: insights from post-soviet Sakha villages. Arctic 59(3):294–310

Crate SA (2006b) Cows, kin and globalization: an ethnography of sustainability. Alta Mira Press, Walnut Creek

Crate SA (2008) Gone the Bull of winter: grappling with the cultural implications of and Anthropology's role(s) in global climate change. Curr Anthropol 49(4):569–595

Crate SA (2011a) A political ecology of water in mind: attributing perceptions in the era of global climate change. Weather Clim Soc 30(3):148–164

Crate SA (2011b) Climate and cosmology: exploring Sakha belief and the local effects of unprecedented change in North-Eastern Siberia, Russia. In: Gerten D (ed) Religion in environmental and climate change: suffering, values, lifestyles. Continuum Books, London, pp 175–199

Crate SA, Fedorov A (2013a) A methodological model for exchanging local and scientific climate change knowledge in Northeastern Siberia. Arctic 66(3):338–350

Crate SA, Fedorov A (2013b) Alamai Tiin: Buluu Ulustarigar Klimat Ularitigar uonna Atin Kihalghar (Alamai Tiin: climate change and other change in the Viliui Regons). Bichik, Yakutsk

Ermolaev V (1991) Uu dolgun kuttaakhtara: agro-etnograficheskai ocherk [Phases of Yakut (Sakha) Land Tenure]. Sakha Book Publishers, Yakutsk

Fedorov A, Konstantinov P (2008) Recent changes in ground temperature and the effect on permafrost landscapes in Central Yakutia. In: Ninth international conference on permafrost, vol 1. Institute of Northern Engineering & University of Alaska, Fairbanks, pp 433–438

Fedorov A, Konstantinov P (2009) Response of permafrost landscapes of Central Yakutia to current changes in climate and anthropogenic impacts. Geogr Nat Resour 30(2):146–150

Netting RMC (1968) Hill farmers of Nigeria: cultural ecology of the Kofyar of the Jos Plateau. University of Washington Press, Seattle

Netting RMC (1993) Smallholders, householders: farm families and the ecology of intensive, sustainable agriculture. Stanford University Press, Stanford

Nogovitzin DD (1985) Vodnye Resursy Iakutii i Ikh Ispol'sovanie [The Water Resources of Yakutia and Their Use]. Siberian Russian Academy of Sciences, Yakutsk

Rosen A (2007) Civilizing climate: social responses to climate change in the ancient near east. Alta Mira, Walnut Creek

Roshydromet (2008) Assessment report on climate change and its consequences in the Russian Federation. ROSHYDROMET, Moscow

Shadrin AP ed. (1984) Antropogennoe Vozdeistvie na Vodnye Resursy Iakutii [Anthropogenic Influences on the Water Resources of Yakutia]. Siberian Russian Academy of Sciences, Yakutsk

Steward JH (1955) Theory of culture change. University of Illinois Press, Urbana

Strang V (2004) The meaning of water. Berg, Oxford

Chapter 2
Sakha Republic (Yakutia): Local Projections of Climate Changes and Adaptation Problems of Indigenous Peoples

Vanda Ignat'eva

The climate is crawling like a snail, barely noticeable, and destroying civilizations – A. Nikonov

Abstract The most dramatic collisions, directly or indirectly related to changes in the global climate, have occurred in Yakutia (Sakha Republic, Russian Federation) through a noticeable growth in natural disasters (such as floods, fires, and drought). The range and scope of these collisions increases the costs of mitigation and continual deterioration of the environment. As sociological studies show, the issues of ecology and environmental change are always present in the list of the stressful subjects, causing serious concerns for the indigenous peoples, together with such vital issues as unemployment and rising commodity prices.

In this regard, it is interesting to consider the local aspects of climate variability in the light of observations, opinions and assessments of the indigenous people themselves, most of whom live in rural areas. This chapter presents the results of a field study conducted by the author in the village of Betenkes, Verkhoyansk Ulus (District). The results indicate that for inhabitants of that transpolar village, climate change is not a distant prospect but a directly experienced reality to which they are currently trying to adapt.

Keywords Climate change • Flood • Indigenous peoples • Security risks to life • Adaptation problems

V. Ignat'eva (✉)
The Institute for Humanities Research and Indigenous Studies of the North of the Siberian Branch of the Russian Academy of Sciences, 1 Petrovskaya str, Yakutsk 677027, Russia
e-mail: V_Ignat@mail.ru

T. Hiyama, H. Takakura (eds.), *Global Warming and Human - Nature Dimension in Northern Eurasia*, Global Environmental Studies,
DOI 10.1007/978-981-10-4648-3_2

2.1 Introduction

It is generally recognized that the greatest threat to the population and economy of Yakutia are emergencies caused by water during spring and summer floods on the rivers. Indeed, despite preventive measures by the government of the Republic of Sakha, there has been an interannual increase not only in the number of those affected by such natural disasters but in economic damage they have caused. That damage involves the evacuation of people, flooding of settlements, meadows and pastures, restoration of roads, and residential and industrial buildings. (Boyakova et al. 2010a, b).

Such negative effects are attributable to a large part of the natural environment of Yakutia having a high probability of flooding. The current settlement system increases the risk of flooding villages, which are mostly within the basins of large and small rivers. According to experts, the worsening of floods along the rivers of Yakutia is inevitable, and is contemporaneous with the increasing effects of global climate change (Arzhakova 2006a, b; Spektor 2006). In this context, research interest lies in the prevention and protection of rural settlements from the negative effects of flood waters.

In the present study, I used field data including the results of a survey of local residents, interviews with key informants (long-term residents, regional ethnographers, folk meteorologists, administration staff), district newspaper publications, and illustrative data. With these data, I outline local projections of climate change, identify existing and potential security risks to livelihoods of the rural population of Yakutia.

2.2 Betenkes: Brief Description of the Study Area

The village of Betenkes (135°34′15″E, 67°38′57″N) is a typical Yakut village, the center of Adychinsky Nasleg, situated on the left bank of the river Adycha, a tributary of the Yana (872 km). The Yana is one of the largest northern navigable rivers of Yakutia. The Yana upstream is one of the poles of cold in the Northern Hemisphere, where the absolute temperature minimum in the world was recorded at −71.2 °C (Fig. 2.1).

The Adycha flows into the Yana 648 km from its mouth. The Adycha's length is 715 km and it has a basin area of 89,800 sq. km. It originates from the western slopes of the Chersky Range and flows within a wide valley. Its water supply is from both snow and rain. The average flow rate is 485 m^3/s. The river freezes in October and thaws in late May. The winter freeze is complete, reaching the bottom for 1 to 4.5 months, forming huge ice mounds. The main tributaries on the right are the Delakag, Charky and Tuostakh, and the Derbeke, Nelgese and Borulakh on the left. The nature of the water regime according to the classification of B.D. Zaykov makes the river a Far Eastern type. The characteristic feature of such rivers is an excess of maximum levels and water flow rates during rain-induced flood peaks in spring. Therefore, the formation of floods is mainly due to prolonged rains (Yakutia 2007a).

Fig. 2.1 Adycha River in Betenkes

The village of Betenkes is part of Verkhoyansk Ulus (District), one of the hardest-to-reach regions of Yakutia. External communications throughout the year are solely by air. Other means of transportation are seasonal in nature and function. These are the river in summer and autumn (late May through September), and automobile in winter and spring (mid-November through April).

The entire territory of Adychinsky Nasleg is within the zone of permanently frozen ground (permafrost). The climate is subarctic, sharply continental, characterized by a strong influence of cyclones and climatological fronts from the north, owing to a lack of natural barriers to winds off the Arctic Ocean. This climate is characterized by a lack of snow winter and warm summers. The average air temperature ranges from −38 to −48 °C in the coldest month (January) to 15–17 °C in the warmest month (July). Average annual rainfall is 150–300 mm (Yakutia 2007b).

In 2010, there were 872 inhabitants of Betenkes, including 304 children under 18 years. According to the local administration, the permanent population is decreasing significantly, associated with frequent migrations within and outside the ulus. The main reason for this movement is the socioeconomic appeal of the ulus center and urban settlements (relative to rural ones), narrowing of the agricultural labor market, and rising unemployment. Because of the migration, the working age population has declined from 460 in 2008 to 382 in 2010.

The basis of the Betenkes village economy, the former central farmstead of the state farm *Adychinsky*, is the agricultural sector constituted by traditional indigenous sectors in Yakutia, i.e., horse (889 head) and cattle (250 head) breeding. In addition, 240 of the 294 families have a personal subsistence plot. In the village, there is the V.S. Chirikov secondary school with a paleontological and ethnographic museum, a kindergarten, a feldsher-midwife station, the House of Culture, trade outlets, a *Post of Russia* department, and a river pier.

2.3 Various Aspects of Climate Variability in the Light of Observations, Opinions and Estimates of Yakutia Rural People

Primary data was collected during personal interviews with respondents. The demographic characteristics of the respondents interviewed were as follows: men 48.5%, women 51.5%; age groups in descending order – young (18–24 years, 24.2%), middle age (35–49 years 27.3%), 14–17 years 15.2%, 50–59 years 12.1%, 60 years and older 12.1%, 25–34 years 9.1%. Ethnically, most of the respondents were Yakuts (97%). In terms of socio-professional status, most numerous were persons with general secondary (42.4%) and specialized secondary education (36.4%), followed by junior secondary education at 15.2% and higher education 6.1%. Occupations were: employed 42.4%, students 21.2%, temporarily unemployed 18.2%, pensioners 12.1%, other 6.1%.

The general background of the sociological study was characterized as positive; most Betenkes villagers showed sufficient understanding of the purpose of the scientific research conducted by the author. Administration employees of Adychinsky Nasleg, many respondents, and key informants expressed their gratitude for the attention to such a complex and controversial issue as the consideration of the effects of climatic factors on the livelihoods of rural people in Yakutia.

According to experts, the mood and behavior of both individuals and groups often depend on their feelings with respect to events and processes they project into the future. Some social movements are even sanctioned by institutional fears (Matveeva and Schlapentokh 2000). Considering this, the respondents were first asked to indicate the problems that caused their most serious concern and anxiety at the survey time.

The list of urgent life problems was headed by a progressive increase in prices (78.8% of respondents) and the frightening prospect of poverty, depriving many people of confidence in the future. This was followed by a theme related to the increase in natural disasters (60.6% of respondents). In the context of my research, I think it very disturbing when in the public mind the protection of villages and their populations against natural disasters (floods and forest fires) is seen as significantly more serious than even issues vital to rural society such as alcoholism (51.5% of respondents), rural unemployment (42.4%), and fear for the material wellbeing of the family (39.4%). The resulting hierarchy of "fears," in my opinion, is an important indication of the strong dependence of human life in the countryside on environmental change, especially the interruption and intensification of the water cycle.

The study found that most respondents recognize the reality of climate change. In particular, 87.9% of respondents agreed with the statement, "*Global climate change is becoming a reality, as evidenced by the ongoing process of a warming world*". The opposite statement, "*There are no global changes in the world climate, as the whole history of the Earth shows that warmer periods are followed by colder ones*," was supported only by 12.1% of respondents. The perspectives of the first group of respondents were supported by personal experience; they stressed that they personally sense climate changes.

Answers to the question "What factors do you think influence global climate change" helped to identify common (according to the public opinion) causes of climate fluctuations. Nearly a third of respondents (27.3%) were convinced that the cause of climate variability is "*natural cyclical fluctuations in the Earth's climate*". However, numerically dominant were respondents that linked ongoing climate change with increasing anthropogenic interference with the environment. In particular, among the eight arguments offered to choose from, the most supported were three: "*cutting down forests, which play an important role in the gas balance of the atmosphere and climate regulation of the planet*" (57.6%); *the growing extraction of fossil fuels and other natural resources of the Earth* (42.4%); *the widespread construction of industrial complexes, damming major rivers of the world*" (42.4%).

In this regard, it is appropriate to point out that formation of *basic knowledge and understanding of climate change* for most respondents (84.8%) was mainly acquired through the mass media, television, newspapers, magazines, and the Internet. For only 21.3% of respondents, *the primary source of information* on climate change was personal observations and informal contacts (stories and other people's conversations). These data support the viewpoint of anthropologists regarding a global transit of information and "mediatization of knowledge," despite the geographic remoteness of the regions of their field work. In particular, Beth Marino and Peter Schweitzer, who studied climate change in five Inupiaq settlements of northwestern Alaska, found that their respondents gave them general information obtained via journalistic, popular scientific, and other information sources. However, when asked about changes in the context of the local environment, they shared their personal experiences based on their daily or seasonal activities. As a result, the authors concluded their work with the suggestion that perhaps the anthropological study of climate change will progress much further if "we stop talking about it" (Marino and Schweitzer 2009).

Despite the popularity of universal concepts of global climate change among the locals, I have also revealed among many of them the presence of individual cognitive experience, based on personal observations of changes in their environment over the past 10 years. They are not only noticing changes in the local ecosystem but are also able to provide a variety of their manifestations in everyday life.

Examples from personal observations of the respondents helped to delineate such fundamental characteristics of contemporary climate and weather in the Verkhoyansk region. These include increased incidence of extreme hydrologic events such as floods and summer overflow (63.6% of respondents), dramatic changes in weather and air temperature (54.5%), and deterioration and loss of forests by frequent fires (51.5%). As indicators of climate variability, they also pointed to melting of the permafrost and bogging of areas, shifting seasons ("*early spring,*" "*prolonged autumn*"), changes in seasonal temperature ("*warm winter,*" "*cool summer*"), average annual rainfall ("*little snow,*" "*a lot of rain*") and the surrounding landscape ("*swamping of the area*"), plus degradation of the animal world ("*animals dying of unknown causes,*" "*reduction of individual populations of animals, birds and fish,*" "*increase in the number of blood-sucking insects*").

The literature indicates that most of the permafrost zone in Yakutia contains rocks with high ice content, which produce the weak stability of northern permafrost complexes. Therefore, the increase in air temperature is associated with the development of such adverse cryogenic events as an increase in humidity of soil and its deterioration, and the degradation of ground ice. Processes in the upper shell of permafrost, particularly a significant increase in depth of the active layer, cause changes in terrain and topography, increased erosion and drought, waterlogging and deforestation (Vasil'ev 2005) (Figs. 2.2 and 2.3).

During the field work, I met an extraordinary personality, a former teacher of physics, local historian, and long-time resident of the village, N.G. Bozhedonov (born 1932). In his diary of instrumental observations, he reflected on the dynamics of an indicator that has the most significant impact on the cryosphere – higher seasonal air temperature. It is noteworthy that in the list of environmental changes, the key informants were changes in landscape. These are the deterioration of forest soils, formation of gullies, erosion of river banks, riparian forest destruction by landslides on thawing slopes, melting of glaciers and snow cover in the mountains between the Verkhoyansk and Chersky ranges, and the emergence and acceleration of "strong wind." The latter is among the most dangerous and unpredictable meteorological phenomena.

For example, ongoing erosion on the coast and flooding and rotting wood in the river Adycha (the main waterway of Betenkes) were reported by the head of the local administration M.I. Osipov:

Fig. 2.2 An indigenous meteorologist N.G. Bozhedonov

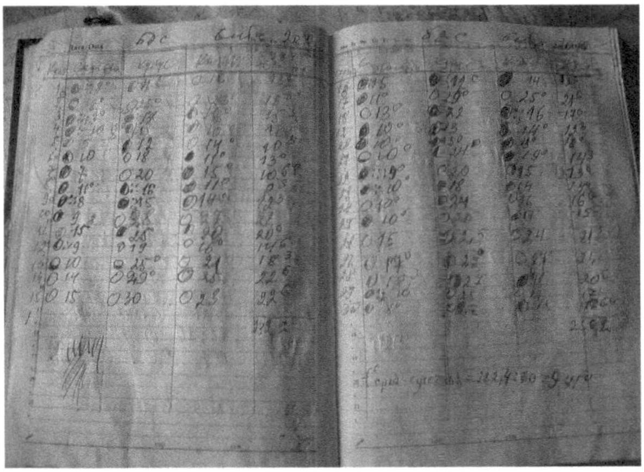

Fig. 2.3 The record of monitoring since 1975 by N.G. Bozhedonov

The village is gradually losing its riparian lands. Due to the constant caving-in of the river banks the people are forced to move their homes and outbuildings. In view of this we will have to develop a new 'Master plan of the village of Betenkes', that will be approved in 2012 (field notes, August 2010).

Information about the melting of permafrost was provided by local hunter N.N. Chirikov (born 1949). With the experienced eye of an indigenous climatologist, Nikolay Nikolayevich detected and evaluated a number of negative effects:

In recent years, the sun has begun to grow hotter, sometimes buluus (the Yakut word for permafrost) has already started to melt. They visited various places and were surprised at how one or another area had changed – in one place gullies formed, in another, on the contrary, soil had bulged and in yet another there was frost that melts only in the autumn, and in some other place a small lake had disappeared. In the locality of Baldyma (450 km from Betenkes) felled trees lie in stacks, and the forest around is completely swamped. I personally observed a phenomenon resembling some gas emission in the area of Achchygyi Uruye (100 km) and on the River Tuostaakh (60 km). All we have seen with our own eyes is the evidence of climate warming. (field notes, August 2010) (Figs. 2.4 and 2.5).

Local residents also reported the addition of bird fauna in the Verkhoyansk region with non-endemic species such as the lapwing (distribution from the Atlantic to Pacific, south of the Arctic Circle) and gull (common in the central Eurasian continent). In their view, expansion of the northern border of natural habitat for avifauna representatives of more southern regions of the earth demonstrates a response of wildlife to climate warming.

It is clear that one cannot assert a direct causal link between climate change and the examples given by my respondents. However, one cannot completely deny the obvious parallels between climatic shifts and increases in frequency and magnitude of floods, as well as adverse effects on local ecosystems worldwide.

Fig. 2.4 Process of gully formation in the taiga area where horses graze (locality of Buordaakh Khaya)

Fig. 2.5 Destruction of coastal forests by landslides on thawing slopes (the area Suus Kustaakh)

2.4 Extreme Floods and Local Case of Climate Change Adaptation

In recent decades, according to N.G. Bozhedonov's observations, floods have become more frequent and extensive. He says:

> *The residents of our village know too well what it means to live under the constant risk of flooding. The risk is at its maximum during the period of spring and summer floods on the river Adycha. Flood is a natural process necessary to the health of the catchment, but it becomes more and more destructive and costly. The critical level of water on the river Adycha for Betenkes is the mark of 12 meters. To my knowledge, in the period from 1943 to 2010, there were 15 major "motuoks"* ("motuok" or "motuokh" is a word for flood in the local dialect, from the Russian word *potop* for flood. – V.I.), *that caused great damage to our village. Note that when the water level is at 11.80 m it floods half of the village, so you can imagine the consequences of the worst flood in the history of Betenkes when the water level rose to 14 meters. It surpassed all the previous floods in the area of inundation, length, volume and height of the flood waters* (field notes, August 2010) (Table 2.1).

The latest devastating floods in Betenkes were in late July 2008, attributable to record heavy rainfall in the Verkhoyansk Mountains. This caused above normal rates of mountain snowmelt and soil water content. Moreover, a sharp rise in the level of the river Adycha was caused by the complex geomorphological structure of the riverbed and, paradoxical as it seems, the flood defenses. These constrained water to the river and increased the flood crest. As a result, water broke through the dam and spread rapidly through the village. The flood zone covered 187 houses and all public buildings. There were 195 people, including 112 children, that had to be evacuated (Fig. 2.6).

The head of the local administration M.I. Osipov bitterly recalls that "*Enormous damage was caused not only to property, but also to the rural infrastructure, in particular, it washed away roads, streets, driveways, and other communication lines. Besides, the water ruined the outbuildings, the stocks of wood and hay. But, most importantly, it killed the livestock (horses and cows), which did not have time to withdraw to a safe place, the main livelihood of rural people. The losses we suffered are just catastrophic*" (field notes, August 2010).

The increased likelihood of extreme meteorological conditions and threat of heavy flood recurrence force many people that have too little money to maintain their standard of living to spend money and labor on adaptation to climatic variations.

Table 2.1 Maximum water level on river Adycha by year

Year	Water level (m)
1967	13.57
1968	13.28
1974	11.90
1981	11.81
1996	13.34
2008	14.00

Data from N.G. Bozhedonov

Fig. 2.6 A unique type of Yakut horse breed, the Verkhoyansk horse, which are bred only in the arctic Verkhoyansk District. During the 2008 floods, more than a hundred horses died in Adycha Nasleg.

Fig. 2.7 Houses on piles in village of Betenkes

This is illustrated by the construction of houses on wooden piles. Despite the high risk of flooding, Betenkes has an underdeveloped infrastructure for flood protection. For example, there is no weather monitoring or early warning systems because there is no meteorological station and watchtower. Works toward stabilization of the riverbed and banks of the Adycha and dam construction are not carried out owing to lack of funding (Figs. 2.7 and 2.8).

Fig. 2.8 The magnified image of the piles of house, an illustration of how villagers avoided the threat of floodwaters

Therefore, adaptation to the increased risk of flooding is reduced in the study area to the construction of houses, whose uniqueness consists of them being built on large wooden piles, submerged into the earth 2.5 m. One pile costs 800 rubles and, for construction of a pile field for a medium-size house, there should be about 40 piles. In addition, processing of the piles themselves costs approximately 12,000–13,000 rubles. Therefore, the sum is about 45,000 rubles. The government of Yakutia compensates households affected by flood 20,000 rubles, plus 2000 rubles for each family member. It is clear that such construction increases economic inequality at the level of rural households, because some rural families have limited financial possibilities.

According to the Regional Department of the Federal State Statistics Service for the Sakha Republic (Yakutia) the list of uluses of Yakutia with a minimum of per capita income (PCI), cost of living (CL), and real cash accumulation (RCA) include rural uluses, inhabited mainly by indigenous peoples of the republic (Statistical Yearbook 2009). In view of this circumstance, it is logical to conclude that in the absence of public investment in infrastructure protecting from floods, income inequality contributes to inadequate protection of poor and non-competitive groups in the rural community from systematic risks associated with climate change.

Fig. 2.9 Construction of the paleontological and ethnographic museum at Adycha secondary school, where the new timber is used for external look

Regarding the issue in question, it is important to draw attention to the lack of funding and logistical support to the social infrastructure of Betenkes. In this connection, M.I. Osipov reported the following:

> *The full recovery of the various facilities intended to provide the working conditions, social and cultural life of the residents of our village, require both federal and republican investment. But we do not see any tangible results here. Therefore, we are forced to look for different ways to solve this problem. Thus, the construction of a new building for our paleontological and ethnographical museum that possesses the largest fossil collection in the republic and is the pride not only of the residents of Betenkes, but of all the Verkhoyansk ulus, is being sponsored by our countryman, entrepreneur V. Kh. Chirikov. The construction itself is carried out by the villagers, but we are experiencing persistent difficulties with building materials. You understand, the way to the polar ulus is quite long—the barge from Yakutsk first goes down the Lena to the ocean, and then up the Yana and the Adycha. For this reason, we have to economize on everything. For example, for the outer skin of the museum building we use new building materials, and for the interior old ones, damaged during the recent floods, but still suitable for construction* (field notes, August 2010) (Figs. 2.9 and 2.10).

Therefore, today we can discuss the impact of climate processes on living and working conditions of rural residents, which allows us to include in the agenda a question regarding protection of the ethnoeconomy and cultural property of the world's northernmost horse- and cattle-breeders.

2.5 Risks from Climate Change

Because many people notice and to some extent sense the changes in climate and weather, an important objective of the present study was to identify the range of climate-related risks that could affect the lives of any Yakut village and its residents.

Fig. 2.10 The interior wall under construction by old timber of the paleontological and ethnographic museum which shows the lack of budget. The exterior-interior contrast shows the hope and the reality of villagers

According to the residents of Betenkes, the most vulnerable target of adverse effects of climate change is human health. In this regard, life in the north was originally paired with the impact on people of an entire range of negative factors associated with extreme climatic conditions, and one cannot really ignore the issue of environmental pollution provoked by global warming (Petrova 1996; Petrova et al. 1995). This largely concerns the deterioration of natural reservoirs, which are the main source of drinking water (Table 2.2).

It should be emphasized that the local community very actively raises the issue of unsafe drinking water as a threat to public health. Thus, during the report of the representatives of the executive power of the Republic of Sakha (Yakutia) to the population of the Verkhoyansk Ulus of Adychinsky Sasleg, it was said bluntly:

> *We are forced to drink water from the river Adycha, which contains toxic trace elements of heavy metals. No doubt, the presence in the water of substances such as copper, zinc, mercury, arsenic, etc., negatively affects the health of people. So we Adychinsky Ulus dwellers have constantly been raising the question of urgency in addressing the problem at various meetings with government officials and other senior leaders of the republic. We believe that providing clean water to rural population is your direct duty* (YSIA 2011).

As noted in the literature, the most dangerous factor for the human body in the far north is sharp fluctuations of basic meteorological parameters, i.e., temperature, humidity, atmospheric pressure, precipitation, wind speed and solar radiation (Proceedings of the 13th International Congress of 2006). Chief physician of the

Table 2.2 Distribution of answers to the question: "In your opinion, what will be the most vulnerable in the process of climate change and the manifestation of its negative effects?", in %

	Weak	Not very strong	Very strong	Hard to answer
My place of residence	9.1	18.2	51.5	15.2
Land of agricultural designation	9.1	21.2	45.5	18.2
Bodies of water	15.2	33.3	36.4	12.1
Forest	12.1	18.2	51.5	12.1
Wildlife	3.0	15.2	**66.7**	9.1
Plants	9.1	39.4	33.3	12.1
Reindeer- and cattle-breeding	6.1	21.2	51.5	15.2
Hunting and fishing	12.1	21.2	51.5	9.1
Folk arts and crafts	15.2	48.5	21.2	9.1
Ecological conditions	9.1	18.2	57.6	9.1
Human health	3.0	9.1	**72.7**	9.1
Drinking water	3.0	12.1	**66.7**	12.1
Culture and language of my people	27.3	30.3	21.2	15.2
Rural population	9.1	24.2	51.5	9.1

feldsher-midwife station in Betenkes, A.K. Moldokunov, spoke about the prevalence of hypertension among the locals, one of the main causes of which he believes to be sudden changes of temperature and atmospheric pressure. He also noted correlation between observed changes in climate and the growth of cardiovascular disease, as well as increased meteopathic reactions, even in healthy people (field notes, August 2010).

During the field work, it became clear that the community is seriously concerned about the potential impact of space weather on human health. Perhaps this concern has something to do with 2009 local in-situ measurements of natural background radiation and electromagnetic fields, the results of which revealed ultra-low frequency emissions in the Verkhoyansk region. These emissions are interpreted as discharges into the ionosphere, called sprites and elves.

Among the potential risks to the daily life and work of rural residents related to climatic factors, the respondents also mentioned ecological problems and worsening conditions of cattle, horses, game animal habitats, birds, and fish, which can harm livestock, hunting, and fishing. Along the same lines are dramatic changes of the natural landscape shell of local ecosystems, which are associated with a reduction in area of agricultural land, above all, of hayfields and pastures along the banks of rivers and streams. The respondents also included in the list of perceived risks the bogging of areas, soil erosion, and forest degradation, which together can be a trigger of negative social and economic processes.

It is significant that many of the respondents, from 14-year-olds through the elderly, share a perspective regarding a strong dependence of traditional activities on climatic fluctuations. Those activities form the basis of their livelihoods. Most of these people are well aware that in the case of the prolongation or increase of climate change, the threat of declining agriculture will become very real. Clearly, this will result not only in gradual mass poverty of the rural areas, villages and local residents but also in distortion of the original indigenous economy. First, this is because climate change in the environment greatly exacerbates existing social, economic and environmental problems of the indigenous peoples of Yakutia (Ignat'eva 2011; Boyakova et al. 2011). Second, the limited material, financial, technological, investment, intellectual and other resources of rural settlements in the republic make their population extremely dependent on external aid. Thus, the issue of vulnerability and adaptation of rural communities and the agrarian economy of Yakutia to climate change has a significant social and ethnic emphasis.

During the fifth summit of leaders of the indigenous people of the Arctic region, *Industrial Development in the Arctic under the conditions of Climate Change – a new Challenge for Indigenous Peoples* (Moscow 2010), representatives of the Association of Indigenous Peoples of the North, Siberia and Far East of the Russian Federation raised the issue that the rights of the indigenous people of the country must be supplemented with special rights, because they first experience impacts related to changing climatic conditions. Climate change contributes to distortion of their original ethnic culture and the economy and, ultimately, to their moving away from traditional ways of life and loss of their ethnic identity. The representatives stressed the importance of developing indigenous climatic strategies, the basic concept of which is that the natives not only claim their rights but attempt to find solutions to the climate problem, in partnership with other groups in Russian society. In this case, monitoring, education, and partnership efforts in the development of adaptation measures and their legislative consolidation are considered innovations in the aboriginal approach to climate change (http://www.arcticpeoples.org).

In light of the foregoing, it is appropriate to refer to the UNDP Human Development Report (2007/2008). It sets out five key risk factors in the development of mankind as caused by climatic changes. These are as follows: (1) Reduction in agricultural productivity, which can increase the number of people suffering from acute malnutrition; (2) growth of water insecurity and increase in the number of people suffering from water shortages; (3) heightened risk of coastal flooding and extreme weather events, which would produce flooding of small island states and increase the number of people affected by floods and the large-scale displacement of these people; (4) collapse of ecosystems: 20–30% of species would face a high risk of extinction in case of excess of additional warming of 3 °C; (5) increasing the threat to human health and risk of disease.

The frightening coincidence of the results of studies by the UNDP with the opinions and assessments of the respondents in Betenkes suggests that earth's climate changes (primarily global atmospheric change and its adverse effects) have become one of the most important problems of the modern world. In this regard, experts on the Intergovernmental Panel on Climate Change (IPCC) stated that if we waste time and do not take necessary steps now, in 15–20 years when irreversible processes have occurred in nature, adapting to the weather "on the move" will "eat" a great deal of money, up to 50% of GNP. Most importantly, it will lead to the loss of human lives.

It appears that the above facts dictate the need for authorities and the citizens of the republic to recognize the reality of climate shifts and, accordingly, the need for development of a climate policy of Yakutia (above all, the policy's key concepts and principles). Given that without an appropriate social basis it is impossible to conduct any policy, a survey was established to identify potential support for a climate policy. The survey revealed that a group of "optimistic respondents" ("*Man is part of nature, and therefore must adapt to all of its changes*") was much larger than one of "pessimistic respondents" ("*Man is unable to oppose nature; periodic climate change will always occur, so these phenomena should be accepted*"). In my opinion, this difference in size between the skeptical (21.2%) and optimistic (78.8%) groups demonstrates clear prospects for climate issues as a specific political resource of power relations.

During my fieldwork, I became increasingly convinced that all my respondents and key informants had in common a great sense of dignity, amazing zest for life, and positive life attitude. Their upbeat spirit and optimism can be expressed by the words of T.G. Potapova (born 1940), a long-time village resident:

> *I and my family, my neighbors, and all the villagers accept any phenomena of nature—both good and bad. Verkhoyansk people have never been coddled by the northern weather, but our ancestors managed to adapt to life on the permafrost. Therefore, none of us is making a disease of the recurring natural disasters (in Sakha "yaryy onostubappyt"); is it possible to feel anger towards the river—our source of water and at the same time perennial threat of flooding?* (field notes, August 2010).

It is significant that the "climate skepticism" current in Russian society was shared only by a small portion (18.2%) of respondents. These people concurred with the potential loss of the usual places of residence. They suggested "thinking over a plan of action for the transfer of human settlements to more favorable places" and "searching without delay for other types of employment for rural people." Even in the event of the worsening adverse effects associated with climate change, the majority of residents (63.6%) intended to actively adapt to them. This finding demonstrates the stability of the concept of adapting to climate-driven changes in working and living conditions.

2.6 Summary and Conclusions

This chapter described results of field work in the Arctic village of Betenkes. It showed that questions of climate risk management are core issues around which other equally important problems are grouped in relation to the management of rural areas, villages and human lives. The "salvation plan" for climate issues proposed by the survey participants included the following: Public investments in developing an infrastructure for protection against floods; investment in projects to create conditions for the conservation of traditional forms of economy of the indigenous peoples of Yakutia; raising ecological awareness and the ecological culture of society; and dissemination of scientific knowledge regarding climate issues. The participants were inclined to believe that the development of strategies to respond to risks posed by climate must be the prerogative of the state and not of private or public initiatives.

In this regard, we note that the planning of measures to adapt to climate change in foreign countries is within the sphere of interest and activities of national and regional governments, local authorities, urban municipalities, insurance companies, donor organizations, and NGOs. These entities have one common goal, to protect people and the economic and social infrastructure from risks posed by climate shocks. In creating adaptation strategies, they use a comprehensive approach that combines government, business, and science, which in the future will reduce the vulnerability of communities and their economies to climate fluctuations. Crucial to combating climate change is an active state policy, and it is becoming increasingly clear that this policy is compatible with sustainable development and economic growth strategies, or is even one of the policy's vital elements. Measures to counter climate change are being developed on both (economic) sectoral and regional (with accounting for geographic, climatic and socioeconomic characteristics of a given area) levels. These measures also provide for the extraction of potential benefits from effects of climate change (http://www.ipcc.ch).

In contrast to developed countries with a wide range of adaptation strategies (from the *UK Climate Impacts Programme* (UKCIP) to the federal Flood Control act in Germany), in Russia there is no uniform state policy on climate change, no national plan for adaptation, and no such plans on the level of individual regions or specific sectors of the economy. The government of the Russian Federation has only prepared a draft *Climate Doctrine* (2009), which provides for improvement of the system of climate observations, climate risk assessment, and development of measures to adapt to climate change.

In this chapter, I have shown specific circumstances related to climate change and social and ethnic overtones of the climatological discourse. From this point of view, I once again emphasize the importance of continued research and public discussion of climate issues, as well as of identifying and assessing the likely impact of climate change on indigenous peoples dependent on their traditional economies, toward developing regional plans for responding to that change.

References

Arzhakova SK (2006a) Conceptual aspects of environmentally sound development of the valley of the Lena River in today's climate. In: Bezopasnots' zhiznedeyatel'nosti v Respublike Sakha (Yakutiya). Materialy respublikanskoi nauchno-prakticheskoi konferentsii (Yakutsk, 30 marta 2006g.), Yakutsk, pp 115–116 (in Russian)

Arzhakova SK (2006b) Floods and safety. In: Bezopasnots' zhiznedeiatel'nosti v Respublike Sakha (Iakutiia). Materialy respublikanskoi nauchno-prakticheskoi konferentsii (Yakutsk, 30 marta 2006g.), Yakutsk, p 116 (in Russian)

Boyakova SI, Ignatyeva VB, Vinokurova LI, Filippova VV (2010a) Yakutia in the context of global climate change: vulnerability, risk, social adaptation. In: Severo-Vostochnyy gymanitarnyy vestnik, №.1 (1), Yakutsk, pp 22–25 (in Russian)

Boyakova SI, Ignatyeva VB, Vinokurova LI, Filippova VV (2010b) Peoples of the Arctic in global climate change: stability, transformation, adaptation. In: Primenenie prirodosberegayushchikh tekhnologiy v usloviyakh kholodnykh regionov. Materialy IX Mezhdunarodnogo simpoziuma po razvitiyu kholodnykh regionov (Yakutsk, 1–5 iyunya 2010 g.), Yakutsk, p 202 (in Russian)

Boyakova SI, Ignatyeva VB, Vinokurova LI, Filippova VV (2011) Social consequences and the adaptation of the population of the Republic Sakha (Yakutia) to natural disasters (based on data from sociological surveys conducted in 2009–2010). In: Severo-Vostochnyy gymanitarnyy vestnik, №.2(3). Yakutsk, pp 37–41 (in Russian)

Fourth Assessment Report, Working Group 1, 2 and 3 (2007) http://www.ipcc.ch

Ignat'eva VB (2011) Problems of sustainable development of the indigenous peoples of Yakutia in the context of climate evolution. In: Tezisy dokladov IX Kongressa etnologov I antropologov Rossii «Kul'turnoe nasledie – resurs innovatsionnogo razvitiya» (Petrozavodsk, 4–8 iyulya 2011 g.), Petrozavodsk, p 261 (in Russian)

In Addressing the Issues of Climate Change Natives Need to Use Every Opportunity. Secretariat of the indigenous people of the Arctic Council (2009) V reshenii voprosov po izmeneniyu klimata aborigeny dolzhny ispol'zovat' vse vozmozhnosti. Sekretariat korennykh narodov Arkticheskogo soveta. http://www.arcticpeoples.org

Marino B, Schweitzer P (2009) Talking and not talking about climate change in Northwestern Alaska. In: Anthropology and climate change: from encounters to action, Walnut Creek, pp 209–217

Matveeva SYA, Schlapentokh VE (2000) Fears in Russia in the past and present. Strakhi v Rossii v proshlom i nastoyashchem, Novosibirsk (in Russian)

Petrova PG (1996) Ecology, adaptation, and health. Ekologiya, adaptatsiya i zdorov'e, Moskow (in Russian)

Petrova PG, Cullberg AY, Volozhin AI, Zakharova FA (1995) Public health and the ecology of the Republic of Sakha (Yakutia). Zdorov'e naseleniya i ekologiya Respubliki Sakha (Yakutiya), Moskow (in Russian)

Proceedings of the 13th International Congress on Circumpolar Health (2006) Materialy 13-go Mezhdunarodnogo Kongressa po pripolyarnoy meditsine (Novosibirsk, 12–17 iyunya 2006 g.), Novosibirsk (in Russian)

Spektor VB (2006) Workshop of the working group of the Northern Forum on floods. In: Nauka i tekhnika v Yakutii, No.1 (10), Yakutsk, pp 32–34 (in Russian)

Statistical Yearbook of the Republic of Sakha (Yakutia): 2009 (2010) Statisticheskiy ezhegodnik Respubliki Sakha (Yakutiya): 2009. Statisticheskiy sbornik, Yakutsk (in Russian)

The Human Development Report 2007/2008. Fighting climate change: human solidarity in a divided world (2007). Doklad o razvitii cheloveka (2007/2008). Bor'ba s izmeneniyami klimata: chelovecheskaya solidarnost' v razdelennom mire, Moskow (in Russian)

Vasil'ev IS (2005) Space-time regularities of the formation of the active layer in the landscapes of Western Yakutia. Prostranstvenno-vremennye zakonomernosti formirovaniya deyate'lnogo sloya v landshaftakh Zapadnoy Yakutii, Novosibirsk (in Russian)

Yakutia: Historical and Cultural Atlas (2007a) Yakutiya: istoriko-kul'turnyy atlas. Moskow, pp 64–65

Yakutia: Historical and Cultural Atlas (2007b) Yakutiya: istoriko-kul'turnyy atlas, pp 804–805

YSIA (2011) Text: «Energetic people of the Adychintsky region». YSIA. Tekst «Adychintsy – narod aktivnyy», 10 fevralya (in Russian)

Chapter 3
Local Perception of River Thaw and Spring Flooding of the Lena River

Hiroki Takakura

Abstract This chapter explores the cultural ecology and feasible methods of adaptation used by the Sakha people in response to spring river flooding. Flooding or overflow occurs during every spring thaw of the Lena River, primarily because of the breakup of ice jams. The Lena flows from the Baikal region in the south and discharges into the Arctic Ocean in the north. Traditionally, the local population or Sakha rural communities in the middle river basin have based their subsistence calendar on the expectation of this natural phenomenon. Recently, the hazard from this spring flooding has increased in both scale and frequency, which in contrast to the traditional benefits of the spring thaw, has had disastrous effects on the local communities. In this paper, I consider the relationship between the human population and the river in the area, and detail indigenous knowledge of the river freezing and thawing and subsistence activities related to this cycle. Sociocultural contexts behind the disastrous flooding are elucidated, and a role for indigenous knowledge in local adaptation policies is advocated.

Keywords Ice-jam flooding • Cultural ecology • Indigenous knowledge • Hazard • Disaster • Lena River

3.1 Introduction

A disaster can be seen as the extreme end of a scale of natural hazards in a particular locale. However, if a certain area is neither populated nor developed, i.e., there are no people or buildings to be affected, we often do not recognize an event in such an area as a disaster. This view is embedded in the sociocultural contexts of human societies, which include our histories, traditions and societal–technological

H. Takakura (✉)
Center for Northeast Asian Studies, Tohoku University,
Kawauchi 41, Aobaku, Sendai 980-8576, Japan
e-mail: hrk@m.tohoku.ac.jp

T. Hiyama, H. Takakura (eds.), *Global Warming and Human - Nature Dimension in Northern Eurasia*, Global Environmental Studies,
DOI 10.1007/978-981-10-4648-3_3

relations. A natural hazard is defined as "an interaction of people and nature governed by the coexistent state of adjustment in the human use system and the state of nature in the natural events system" (White 1974: 4). It can also be seen as a situation in which an extreme natural event exceeds the normal capacity of a human system, which results in a disaster (Oliver-Smith 2009: 13; Reycraft 2000: 3). In this sense, disaster is a sociocultural phenomenon with some degree of continuity within the historical–geographical matrix. The anthropological approach to disaster reveals the local contexts that configure the forms and degree of a particular disaster.

Rather than immediate quantification of the magnitude of a hazard or evaluation of policies and adaptation measures to address a disaster, anthropologists prefer to investigate the sociocultural contexts of these events by asking the following types of questions. Is this really a disaster for the local people? Can local communities develop a method to anticipate and respond to a disastrous situation? Is the local population always vulnerable in a given area? Can cultural backgrounds and social systems of local communities be adaptive and useful in preparation for dealing with an anticipated disaster?

These questions are related to the concept of vulnerability and address social systems such as culture, social institutions, human relationships, and human–nature relationships as a whole. A natural disaster does not affect a given population equally. Rather, it results in a diverse catastrophic experience or risk to each individual (Oliver-Smith 2009: 14). From this viewpoint, a disaster can be the best opportunity for understanding the essence of societal mechanisms such as kin relations, other reciprocal ties, and societal resilience (Hoffman and Oliver-Smith 2002: 14; Tanaka and Hayashi 1998: 14). A disaster also provides context for an integral understanding of how closely related a certain human system is to a particular dynamic of nature, because such catastrophic events are the most extreme point in a given continuum of interactions between humans and nature. This is an important anthropological endeavor, because it can reveal unique relationships between humans and the natural world in which they exist. The anthropology of disaster explores not only sociocultural settings but also recognizes changing natural environments and includes events at the extreme end of the continuum between humans and nature.

Recently, hazards related to repeated spring flooding of the Lena River have increased and caused serious socioeconomic damage to communities along the river and its tributaries (Filippova 2010; PRS 2010; Sukhoborov 2006). The local administration and republic government urgently need to develop effective polices and preventative measures in response to this flooding. The population living in the middle basin of the Lena River is the focus of this paper; this group is known as the Sakha people or the northernmost Turkic people in Eastern Siberia (Crate 2006; Takakura 2009, 2015). Their traditional economy comprises cattle–horse pastoralism, hunting and fishing, and some agriculture. This population has historically lived in riverine valley loci along the Lena River and its tributaries. Here, there is repeated river freezing and thawing in accord with seasonal temperature change, and the natural environment and river thaw inevitably produce spring flooding.

Reasons for recent disastrous effects of this natural phenomenon have not been confirmed. Some natural scientists have focused on recent climate change, showing that trends in global warming correspond closely to acceleration in the scale and frequency of the spring flooding and change in the time of ice breakup in the Arctic and sub-Arctic (Prowse 2007). Russian scientists in meteorology and hydrology regard anthropogenic activity as a very serious factor (Kustatov et al. 2012).

It is not within the purview of anthropologists to investigate cause–effect relationships underpinning extreme flooding. Rather, the objective is to explore local contexts of the spring flooding in relation to livelihoods of the local population and the way in which this flooding becomes a disaster. As discussed above, anthropological disaster study has two orientations. One investigates societal vulnerability and related adaptation policies for protection from disaster. The other explores human–nature interaction on both normal and extreme scales of natural hazards, representing a perspective that sheds light on the local context of human use systems within the dynamics of nature. This paper has the latter orientation in an examination of the Lena spring flooding.

The largest-scale flooding of the Lena River in the past few centuries was in 2001, when the water rise primarily affected Lenski city (Sukhoborov 2006). This instigated the initiation of a prevention program by the federal government, aimed at adapting the city to the threat of floods (Stammler-Gossmann 2012: 68). Although a hydrologic approach to the spring flooding or ice-jam flooding of the Lena River is being developed (Burakov et al. 2007; Kilmjaninov 2007; Kustatov et al. 2012), only a few studies have focused on the human dimension of the flooding.[1] The purpose of this paper is to ethnographically detail local perceptions of the spring flooding, show the extreme scale of such events, and examine the relationship between residents and the river.

How do the local people conceptualize the flooding, regarding its occurrence both as a usual event and extreme episode? What human-environment relationships do the residents conceive in terms of subsistence within the freezing and thawing riverine valley environment? Are both local indigenous knowledge and traditional subsistence activities adaptive to the recent disastrous spring flooding? These questions concern cultural ecology, because this can ascertain whether the adjustments of human societies to their environments require particular modes of behavior or a certain range of possible behavior patterns (Steward 1955: 36). This perspective makes it possible to present the state of sociocultural contexts of local communities that make a particular hazard or a range of natural dynamics result in a disaster or not.

The first task of the present work is to describe local knowledge and subsistence activities in the riverine valley environment and to uncover the local sociocultural contexts of the Lena spring flooding. The second task is to consider the role and effects of the local knowledge and subsistence as related to policies and measures

[1]Anthropological research on climate change and related disasters in the Sakha region has only recently begun. Crate (2008) discusses the anthropological role in climate change research by examining the potential of case studies of the Sakha in the Vilui region and their testimonies regarding local climate changes. Stammler-Gossmann (2012) addressed local perceptions of flooding and the community participation process for local Sakha adaptation in the Tatta region.

aimed at mitigating the flooding disaster. Finally, there must be a discussion of how, or if, these two issues are connected.

The organization of this paper consists of the following four sections. In the second section, I briefly describe the spring flooding in the Lena River system as a recent socioeconomic disaster. Then, I present an ethnographical description of the relationship between the Sakha people and ice of the Lena, including subsistence activities involving the frozen river and adjoining lakes. Subsequently, I describe local perceptions of the flooding and ethnic river hydrology. Finally, I consider the potential of indigenous knowledge and its role in local adaptation policies.

3.2 Spring Flooding as a Disaster

In response to the recent heavy spring flooding of the Lena River, both the government and academics have explored reasons for the rise in water levels and appropriate adaptation policies to prevent further damage (Kilmjaninov 2007; Kustatov et al. 2012; Rozhdesvenskii et al. 2008; Sukhovorov 2006). There are several factors behind these efforts. One is the Russian federal government's concern for establishing preventive adaptation policies against hydrometeorological events (such as flooding) related to climate change (Rakkolainen and Tennberg 2012). Another factor is local concern for transportation. Construction of the Siberian railroad has almost reached the capital city of Yakutsk, and the final stage of this line will involve the planning and construction of a bridge spanning the Lena River. Thus, the seasonal spring flooding should be under government control and management (Rozhdesvenskii et al. 2008: 54).

On March 27, 2010, the government of the Sakha Republic in the Russian Federation adopted a resolution "... on the establishment of the concepts of protection for settlements and economic facilities from the flood". According to the document, the republic is the region with the most substantial severe flooding over the last 12 years in the Russian Federation. Against the background of global and regional climate change, catastrophic floods occur very frequently in this region. The economic damage represents a major loss. Damage from the catastrophe in 2001 reached 20% of the republic's annual revenue. When the damage was not as catastrophic, economic loss ranged from 1.2 to 1.7% of revenue (Table 3.1). In addi-

Table 3.1 The Revenue of Sakha Republic and flooding damage

Mil. ruble/ year	1999	2000	2001	2002	2003	2004	2005	2006	2007	2008
Revenue	16290	24973.2	35610.9	35791.7	39903.9	42888.3	51754	56028	65344	79625
Flood damage	0	0	7000	114.6	0	439	97.4	7.7	1088.5	939.1
%	0.00%	0.00%	19.70%	0.30%	0.00%	1.00%	0.20%	0.00%	1.70%	1.20%

Source: PRS (2010) and Reginony (2004, 2010)

Photo 3.1 Flooded houses viewed from the road being toward Tulagino village from Yakutsk (Yakutsk, May 2010)

tion, the flooding had a negative effect on both the health of the population in the area and the ecology of fauna and flora (PRS 2010). Adaptation policies aimed at disaster prevention and measures for restoration from severe flood damage have become central issues for the republic government.

The spring flooding of the Lena River clearly causes damage to the regional society and economy in general. However, what actually occurs during the spring flooding season at a particular location? During my field research in rural villages of Central Yakutia from May to June 2010, I found myself in the midst of a disastrous season of the Lena spring flood (Photo 3.1).[2] This event was sufficiently important to merit wide coverage in national television broadcasts and made newspaper headlines. No human deaths were reported. However, 10,622 people were affected by the flooding, which reached settlements in 44 areas and killed 1201 head of cattle and 892 horses. Below are details for two villages stricken by the floodwaters.

Residents of Tulagino village in the Yakutsk city special district have had substantial experience with severe spring flooding (Fig. 3.1). The village is very near tributaries of the Lena River. Almost every year, this village is affected by flooding of various degrees. Therefore, the residents are familiar with methods to prevent losses from this flooding. The local administration releases an emergency call to the population on the day prior to when the flood is forecast. The most common measure

[2]Field data were collected during intermittent fieldwork during November 2007, October–November 2008, May–June 2010, and September 2012 in rural villages of the middle Lena River basin.

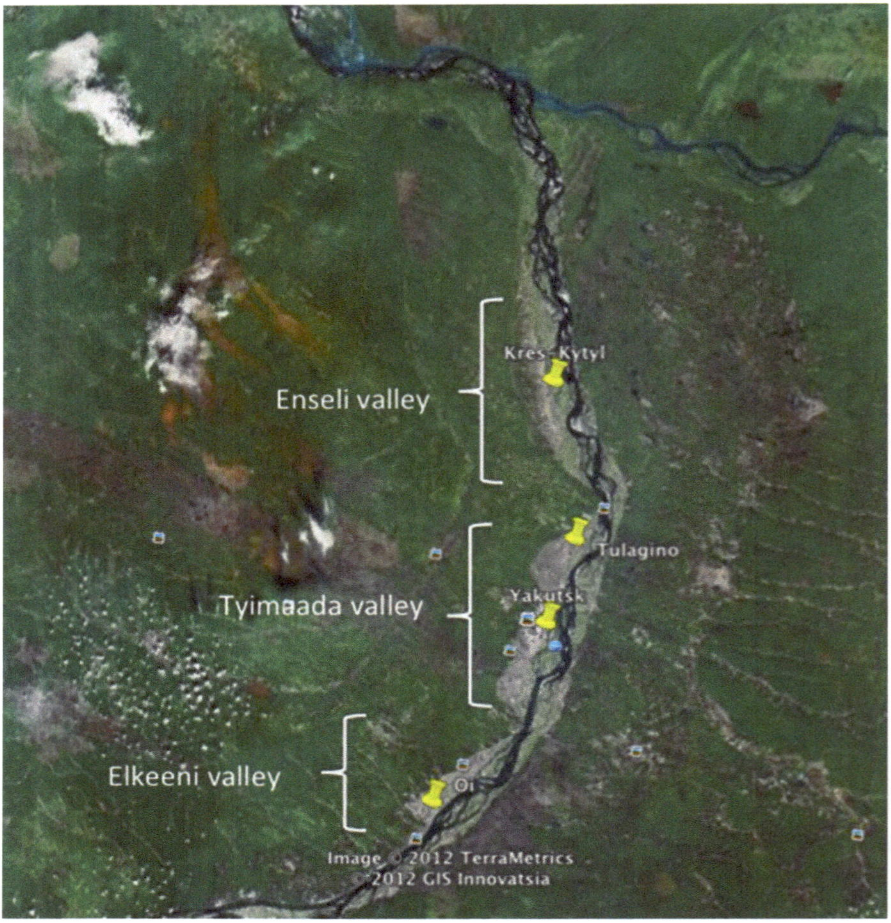

Fig. 3.1 Field research map

is moving personal property to higher locations. Major foodstuffs such as meat, fish, and certain vegetables are kept frozen inside local underground storage spaces such as *buluus*, the local name for the ice on permafrost ground (Photos 3.2 and 3.3). These foodstuffs are brought to the attic spaces of houses, along with electronic devices and computers. Beds, sofas and other items of furniture are put atop tables to avoid rising water. In 2010, the residents finished this preparation during the daylight hours of May 20. Then, during the evening and night, there was a power failure, and they frequently checked river conditions by car and exchanged this information via mobile phone.

The population of Tulagino village was 1502 persons at the time of the flood. The number of homes with flooding reaching above the midlevel of the house was 102 from a total of 415, or 24.5%. It should be emphasized that there was no loss of

Photo 3.2 Frozen meat and potato kept in the permafrost freezer (Tulagino village, May 2010)

Photo 3.3 The transport of foodstuff as preventive action against flooding (Tulagino village, May 2010)

livestock owing to information in the flood forecasts and subsequent appropriate measures taken by the residents. A few days before the flooding, cattle and horses were moved from the river island and other areas subject to flooding to higher forest areas (Photo 3.4).

Photo 3.4 A happy return trip with livestock from evacuation area (Tulagino, May 2010)

In contrast, the people in Oi (Nemiug) village of the Khangalasskii district are not accustomed to catastrophic flooding. Moreover, in 2010, the amount of water here from the spring flood was much greater than in Tulagino village. This resulted in relatively heavy damage, especially to livestock.

At the time of this flood, the population of Oi was 2498 and the village comprised 700 houses. Of these, 209 houses or 29.9% had some flooding. No human deaths were reported, but 96 head of cattle and 51 horses were killed (Photos 3.5 and 3.6). The population uses the riverine valley and island as pasture land, and water overflowed from the branch river to the riverine land and village because of an ice jam that had formed on the Lena mainstream. According to the local administration office, there were no warnings or specific instructions from regional authorities just prior to the flooding. Therefore, that office was unable to warn the population of the danger from the rising waters. The residents of Oi village did not have any experience with such heavy flooding prior to 2010, so they did not know what to expect when it occurred and did not move their livestock from pastures near the river.

The frequency of heavy spring flooding along the Lena has been increasing over the last few decades. When these floods occur, houses and buildings are flooded and livestock pastured on the lower riverine land may be killed. In both cases described above during the 2010 season, 20–30% of houses were flooded, but there were no fatalities in the villages. However, the difference in the loss of livestock was stark. Appropriate forecasts from local authorities and personal experience of the residents with flooding were determinant in preventing damage.

Photo 3.5 Remained ices after the flooding (May 2010, provided by Khangalaskii District administration)

Photo 3.6 Killed livestock by the flooding (May 2010, provided provided by Khangalaskii District administration)

3.3 Subsistence Activities and Interaction with Magnitude of Cold

3.3.1 Valley Natives

Disastrous flooding is an event that can be viewed as an extreme natural hazard, one that exceeds the normal capacity of human systems. To understand that normal capacity, we must ethnographically describe human–nature interaction "governed by the coexistent state of adjustment in the human use system and the state of nature in the natural events system" (White 1974: 4). As shown above, spring flooding of the Lena River is an annual event, but it varies in scale. Therefore, the questions to ask should address the sociocultural contexts of human-use systems in the riverine environment. Here, I explore methods involved in subsistence activities of the Sakha rural population in the study area, and then examine how the people perceive river conditions that may cause disastrous flooding.

The main environmental features of the Central Yakutia are the boreal forest, *alaas*[3] ecology (thermokarst landscape), and the Lena River system with its thousands of tributaries. The *alaas* is a geographically unique topography that has developed in this region, which comprises meadowlands with lakes surrounded by forests. Meadows or open space inside the forest is an object of traditional land use by the Sakha people, as arctic pastoralists (Jordan and Jordan-Bychkov 2001; Takakura 2010). Another area of land used in the same way as the forest pastures and meadows are the river valleys of the Lena Basin. There are three large valleys in the middle basin, the Enseli, Tyimaada, and Elkeeni (Fig. 3.1). Widths of the river valleys can reach 30–40 km (Saito 1985: 71) and the total length of the three valleys is ~180 km. Tyimaada is a well-known valley that is often referred to in an oral tradition with local ethnic origin (Okladnikov 1955: 354). The valley is an important component of local identity for the people. This can be seen by phrases such as "valley native "(*khocho oggoto*) and "*alaas* native" (*alaas oggoto*). Thus, the local lexicon implies that the Sakha are traditional dwellers of the Lena River valley.

How do the valley native people make use of the river environment, and what features are apparent in their subsistence activities? Generally, as pastoralists and hunters, the Sakha use the river floodplain and island as pastures and meadows, or as hunting grounds for geese. They also engage in net fishing along the river. The key issue here is to explore the method of human use systems with regard to the varying state of the river environment, such as spring flooding. First, I detail two unique activities, ice basket fishing and collecting ice blocks for drinking water.[4]

[3] "Alaas" is originated from Sakha language, which is borrowed into Russian as "alas". The alas is usually referred in scientific literature. I use the term "alaas" here because of the cultural implications.

[4] Some ethnographic date in this section were already published in my previous book (Takakura 2015, chapter 4) but I revised them for understanding the local knowledge of river ice process.

Through these descriptions, we can recognize the remarkable human–nature interaction embedded in the arctic river environment of freezes and thaws.

3.3.2 Ice-Basket Fishing

Ethnographic data for the ice-basket fishing were collected by participant observation from October 18 to 24, 2008 in Kres-Kytyl village of the Nam district. I have done fieldwork in this village since 2007, and by chance I had an opportunity to participate in this type of fishing. The village is located in the river valley and has a population of almost 1500. Ice-basket fishing is called *tyy*, and is one of many types of local fishing. The name refers to the way that the basket is sunk deep into the water (Photos 3.7 and 3.8). I observed an older fisherman who was a retired village resident who checked his basket daily. The lake is about a half-hour walk from the village, and the time needed to check a basket is less than a half hour, so the total time required for this fishing is less than 1.5 hours per day. The target is a small fish called *mundu* (*Phoxinus perenurus*). Table 3.2 shows the result of fishing over a week. The average catch was 916 g per day. In the household of the fisherman, which consisted of his wife and a granddaughter attending high school, the catch was used for domestic consumption and not social exchange. In other words, ice-basket fishing provides the subsidiary procurement of protein for this family.

Thus, we see that ice-basket fishing is an individual activity that is not economically significant, but we recognize an intriguing relationship between the residents and natural environment of the river valley. The fishing season is only during early May and late October. The reason why fishing does not continue through the winter

Photo 3.7 Preparation for ice basket fishing (Kres Kytyl, October 2008)

Photo 3.8 Harvest from the ice basket fishing (Kres Kytyl, October 2008)

Table 3.2 Harvest of ice basket fishing

Result	Total catch 5500 g/week, 916 g/day, 3 person-household						
Date	18-Oct	19-Oct	20-Oct	21-Oct	22-Oct	23-Oct	24-Oct
Catch (g)	900	0	2100	–	1000	900	600
Evaluation	Normal	None	Good	–	Normal	Normal	small

season is that the people recognize the limited period for breaking the ice by hand ax. As temperatures fall and the ice thickens, a simple hand ax is not adequate to make fishing holes in the frozen water surface.

The people are very conscious of the freezing season or winter onset, which is in late October through mid-November, because this knowledge is essential to their livelihood. This is represented by expressions such as "when the livestock is frozen soon after the slaughter," "when the ice thickness is between 15 and 20 cm," "when there is not so much snow." I will emphasize this selective usage of degrees representing cold conditions in relation to rural life and subsistence activities in the area.

3.3.3 Collecting Blocks of Ice for Drinking Water

It is important for the Sakha people in the study region to prepare blocks of ice for use as drinking water. Houses in these rural areas usually feature *buluus*, where the temperature can be kept below zero degrees Celsius throughout the year. People use these storage areas as freezers, not only for meat and fish but also for ice blocks.

Ice as a source of drinking water is culturally valued by the local people. It is called "ice-block water – *muus uu*." They prefer it to tap water. One Sakha woman told me that when her mother drinks tea, she could discern whether it was made from ice-block water or tap water.

During the socialist regime, water supply via truck service was introduced, but the local residents continue to prefer water taken from block ice. Today, even private retailers of these ice blocks have appeared in the market economy of the capital city of Yakutsk. Rather than purchase the ice, many rural people collect ice on the lake. As shown in Photos 3.9, 3.10 and 3.11), each household prepares a large amount of this ice for annual consumption.

One case can provide some detail. Field observation was conducted at Oi village on the morning (9:30–12:20) of November 4, 2007. Six kin-related men from five households participated in this activity. The total amount of ice collected was ~14,073 liters. First, I measured the size of the pool that the men made in the lake and then calculated average ice thickness (575 cm * 890 cm * 27.5 cm). From this, it can be calculated that one person requires 2.5 l of water daily, and the amount of ice taken would provide water for fifteen persons annually (14,073/2.5*365). There were 26 members of the participating household. Theoretically, if this activity were repeated twice or three times, the household could satisfy their minimum water annual requirement.

Photo 3.9 Collecting blocks of ice (Oi, November 2007)

Photo 3.10 The lake for the collecting ice blocks (Oi, November 2007)

Photo 3.11 The blocks of ice piled up in a house (Oi, November 2007)

The season for this ice gathering is recognized by the population to be limited to the end of October through the beginning of November. They identify this period as "not-so-thick ice" and "not-so-much snow," which is scientifically accurate. Figure 3.2 shows air temperature change from the past months on average. The most drastic monthly drop in temperature during the year is in November, −23.3 degrees. The end of October through beginning of November immediately precedes this rapid cooling. According to the local people, when of ice-block collection is done after mid-November or the temperature is too cold, the ice is too thick and not readily

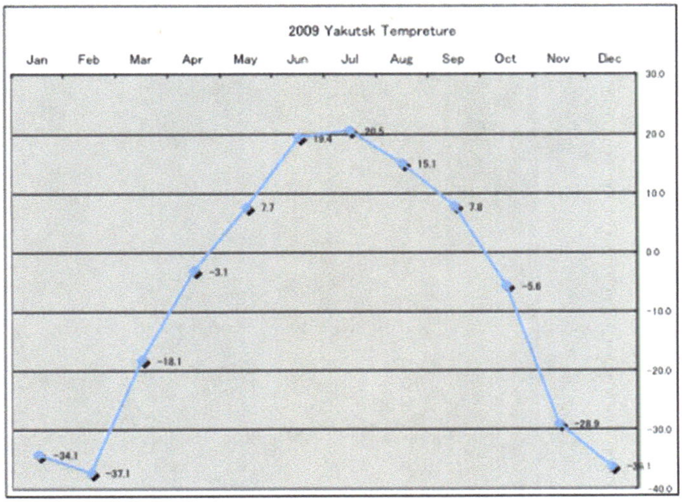

9月	10月	11月	12月	1月	2月	3月	4月	5月
-7.3	-13.4	-23.3	-7.2	+2	-3	+19	+15	+11.7

Fig. 3.2 Mean monthly temperature in Yakutsk

transportable. Before the end of October, the ice is too thin to support people working on its surface.

3.3.4 Disturbance of Nature and Beginning of Winter

The distinctive feature of the two subsistence activities, ice-basket fishing and collecting ice blocks for drinking water, is the selective and intensive use of knowledge on the magnitude of cold in the lake environment. The cold climate in the region makes for a long duration of freezing of rivers and lakes. However, residents carefully distinguish periods when they can manage the ice through indigenous technology and local social organization.

Topographic conditions are another important factor for such activities. As noted above, fishing and ice collection are done on the lake, which is very close to the Lena River. Satellite images of these areas are shown in Figs. 3.3 and 3.4. Although the local people identify both these areas as lakes, their shapes are narrow and long, and they may be isolated tributaries of the river. When spring flooding occurs on an either large or small scale, the water overflows into the riverine lowland embracing

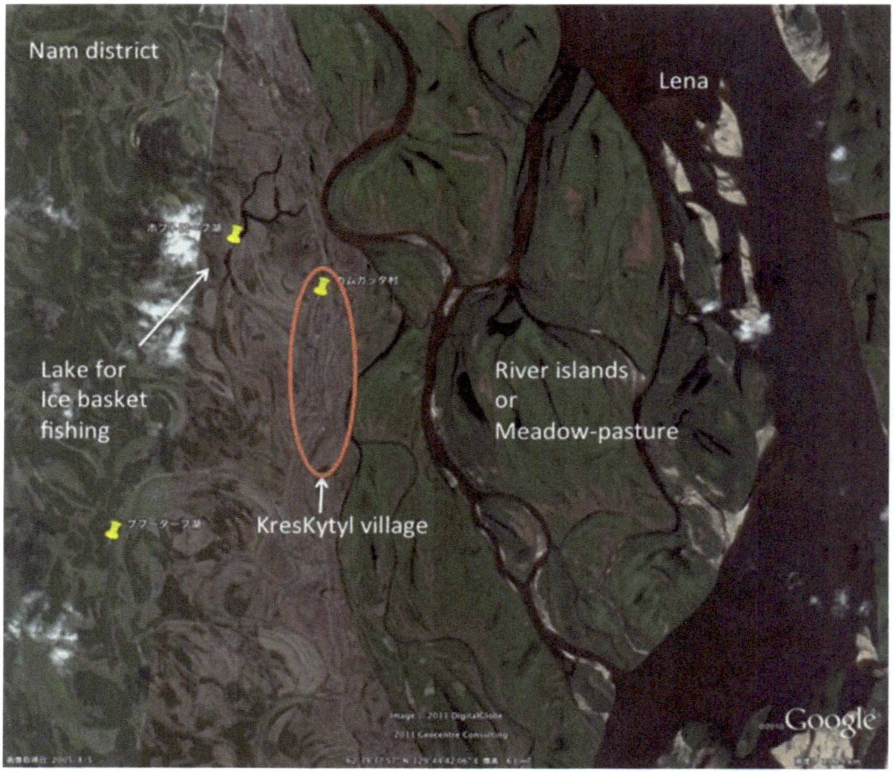

Fig. 3.3 Satellite image of Kres Kytyl village (from Google earth)

these lakes, which revitalizes the lake water and its wealth of fisheries. The local Sakha population also uses the riverine valley as grassland for horse–cattle husbandry and haymaking. Some parts of these lands are subject to private use for haymaking and other parts are local commons for pasture of free-ranging livestock (Takakura 2010).

In general, the annual spring flooding brings sediments that enrich the riverine plain with alluvial soil and contribute to biological production and diversity (Beltaos et al. 2006; Prowse 2007: 210; Yoshikawa et al. 2011: 3). This flooding, a regular natural disturbance, is necessary for revitalization of local fishing and water resources. Some subsistence activities require a certain degree of spring flooding, and this is representative of the human–nature interaction in local communities of the middle Lena Basin.

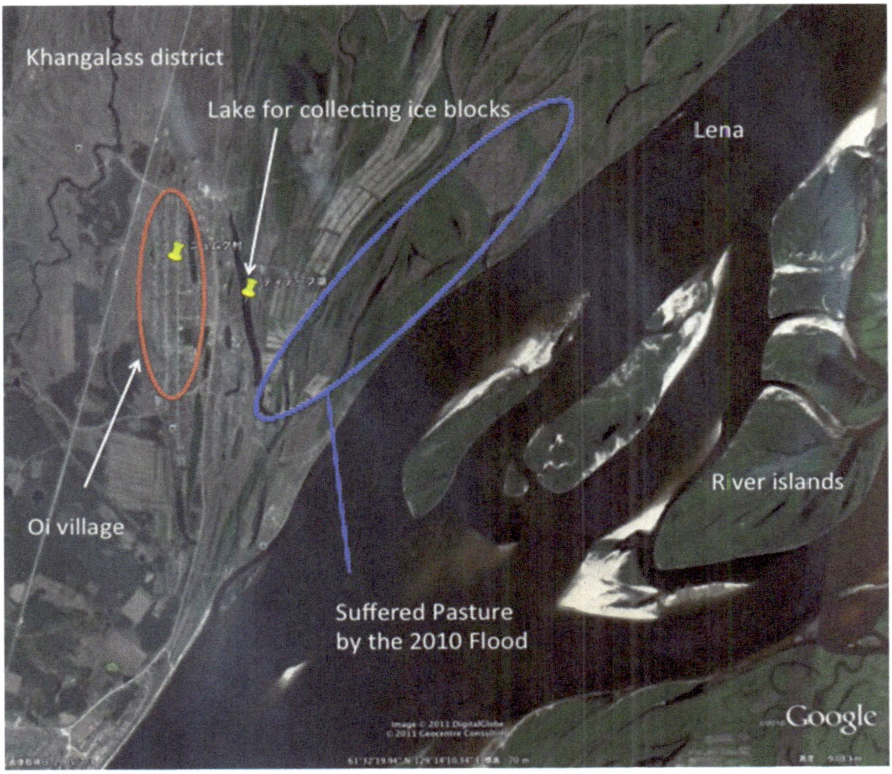

Fig. 3.4 Satellite image of Oi village (From Google earth)

3.4 Local Perception of Spring Flooding

3.4.1 Method of Interview

How do the local people perceive spring flooding along the Lena River? If the flooding has a positive or negative effect on their subsistence activities, social organizations should form a certain range of possible behavior patterns (Steward 1955). I will show some of these behaviors through a description of the way residents conceptualize the freezing and thawing of the river in their language.

Ethnographic data were collected through fieldwork in Tulagino village of the city of Yakutsk, Oi village, and Kres-Kytyl village from 13 to 30 May 2010. I interviewed 29 persons, 22 males and seven females using an unstructured approach. The initial questions asked how river ice floes appeared and how related processes and events occur on the river. These questions were then followed by a period of free conversation. The age range of informants spanned five decades and consisted of three persons born in the 1930s, four in the 1940s, seven in the 1950s, eight in the 1960s, three in the 1970s, and four of an unknown age. Most were rural residents and cited occupations in areas such as farming, private retail, transport and public sector work. Some were pensioners. Some had detailed knowledge about river freezing and thawing, but others did not know much about this cycle. This unequal distribution of knowledge was very apparent. The purpose of this work was not to pursue quantitative analysis or vulnerability through analysis of this unequal distribution, but rather to describe the way in which the people perceive the process of river hydrodynamics and spring flooding. Therefore, I now furnish some representative descriptions of the process and events below.

3.4.2 Indigenous Knowledge of River Freezing and Thawing

One interviewee explained the freezing process of the river in an interesting manner. In this account, first sludge ice or frazil ice (*kyd'ymamakh*) appears in the river during mid-October, and then the sludge ice gradually covers the river surface. Subsequently, that surface becomes completely frozen, a state the people call "river stands" (*oerues turar*). At this point, people can walk on the surface. During this process, on a large river with strong flow like the Lena, frazil ice often forms shapes similar to planks. These formations may rise up and stand vertically on the frozen river surface. There are several different variants of the freezing process according to the informants, although in all of them the key concept is the sludge ice.

The river thaw in spring causes flooding, which is an annual event for the local people. They perceive this process in the following way. During winter, the ice or *muus* covers the river surface and, during spring, the *irii* phenomenon begins, in which snow on the river surface melts. As a result, meltwater persists atop the ice on the river. After this, ice along the riverbanks begins to thaw and then water flows between the riverbanks and the ice cover. This is called *yrbyy* in the native language. When the *yrbyy* appears and develops, it can be difficult for people to walk on the river ice. As the thaw continues, river ice becomes transparent or blue, at which stage it is known as *suraahyn*. In this period, some *koemueoel* or small ice fragments separate from the larger ice cover appear, and water or *uu* flows in open fissures within the river channel. This signals the beginning of floating ice (*muus ustar*). Some of this ice flows and other chunks of it accumulate in certain places, depending on the width and depth of the river and geographic features of the riverbank. This creates ice jams, which are called *kharyy* in the Sakha language. This *kharyy* creates the spring flooding (*uu*).

According to a supplemental interview in Kres-Kytyl village during 2012, *yrbyy* usually appears in early May when the residents set fishing nets on the *yrbyy* riverbanks. The period of *yrbyy* net fishing lasts only about 1 week, because the spring flooding begins rapidly. Interestingly, the *yrbyy* phenomenon is a signal for seasonal migration within the traditional lifestyle of the region. The Sakha are horse–cattle pastoralists and their residence pattern is transhumant; they migrate between summer and winter camps, both featuring log cabins. Some local people had their winter camp on the river island. They had to move off the island to the summer camp on the riverside prior to the appearance of *yrbyy*.

Throughout the research interviews, the river thaw was explained in great detail, although there was some variation in local expressions. This was evident from the vocabulary related to the process. Here, I need to emphasize the local population's definite and specific knowledge of the process of freezing and thawing on the Lena. In particular, the people perceive each phase of the thawing process of river hydrology. This indigenous knowledge appropriately describes reasons for the spring flooding.

3.4.3 Types of Flooding and Pasture

Many informants suggested that the spring flooding is a necessary event for pasture management and haymaking preparation. According to one informant who is an individual cow farmer, "*Kharyy* (ice jam) is necessary for the river island. When it forms near the island, water inevitably appears and then the grass grows well." His meadow for haymaking is on the river island, and his story exemplifies how the ice jam is a necessary natural phenomenon for pasture management in animal husbandry (Fig. 3.5).

Fig. 3.5 River islands as a meadow for haymaking in Lena, a view from the helicopter (Khangalasskii district, October 2007)

There are four types of spring flooding: (1) *khaatyn tolorbut*; (2) *kyra uu*; (3) *orto uu*; (4) *ulahan uu*. The first type is when the river channel is completely full but there is no overflow, the second is small-scale flooding, the third is moderate flooding, and the fourth is extreme flooding. According to the informant, the river does not flood in (1), in (2) and (3) conditions are favorable for the growth of grassland, and (4) can be a disaster.

People also recognize seasonal floods that are different from the aforementioned events. Supposing that the spring flooding is the first stage, there are then several types of seasonal flooding. The second stage is called *khara uu* or "black flooding," which is not beneficial for farming. The reason for the name black flood is that the water comes from snowmelt on soil, not from river ice. This water originates from the Baikal region of the upper Lena River. *Khara uu* is called either second flooding (*ikkis uu*) or flooding with soil (*buordaakh uu*).

Residents explain that the second period of flooding is usually 10 days or 2 weeks after the first period. During that interval, grass usually begins to sprout, so a black flood would damage this growth. However, if the black flood is confined to the valley and island and lasts only a short time, there may be less damage to the pasture. There is a third stage in July, known as summer flooding (*saiyngngy uu*), which is alternatively called grass flooding (*ot uu*) or additional flooding (*kybytyy uu*). In some years, there can also be a fourth flood period, referred to as autumn flooding (*kouhoungngou uu*), in August–September. The third and fourth periods are regarded as harmful to pastures and damaging to haymaking. Local animal husbandry managers need moderate spring flooding for pasture management, but other stages of flooding bring disastrous effects.

3.5 Discussion: Indigenous Knowledge and Disaster

In this paper, I address the sociocultural context of spring flooding on the Lena River. Through the description of ice-basket fishing and collection of ice blocks for drinking water, I explore uniquely feasible methods of sustenance activities that have developed in consonance with the relationship between area residents and ice that forms on the river. The distinctive feature of this relationship is the careful perception and selective use of various magnitudes of cold and related conditions of the river ice process. The residents have also developed the ability to sufficiently recognize the river environment to express in detail various phenomena within the hydrologic process. One may state that their subsistence activities are dependent on the premise of moderate spring flooding. This natural disturbance of the river environment is a necessary element in the Sakha livelihood in relation to their local resource procurement. If it does not reach a severe magnitude, the spring flooding is a benefit to local subsistence activities.

Is the local knowledge of river dynamics adaptive to catastrophic spring flooding? It may be that people with detailed local knowledge can surmise a series of possible events from the onset of river thaw. Generally, the local knowledge simply provides a concept of each condition of ice events and the full framework of the change process. Therefore, the locals can foresee what events may occur during the river thaw and make adequate preparations.

They cannot, however, accurately predict when the water rise will begin and the scale of the flooding. To do this, one would need quantitative data and then be able to examine how these data are interrelated under certain conditions, in order to estimate the scale and timing of a particular hazard. Much more precise information is necessary for the implementation of practical adaptation activities to mitigate disastrous flooding along the Lena.

Nevertheless, I do not believe that the above means that the indigenous knowledge is not useful. As demonstrated in the previous section detailing cases of severe spring flooding, we certainly see that the two villages responded differently to disaster and the results show that these differences had major effects. Tulagino village, which is accustomed to the flooding, prepared well and sustained no damage to livestock. Oi village, which has not had experiences comparable to those of Tulagino, suffered heavy livestock losses. The intellectual framework for recognition of potential events during river thaw can give local people the ability to adapt and prepare for these events.

Consider the fact that the frequency of catastrophic spring floods has recently increased. To look at this phenomenon in another way, the residents' primary experience has been with moderate and small-scale spring flooding, and their indigenous knowledge is organized around these types of events. This may be the reason why local knowledge cannot help the residents predict dates and locations of disastrous floods. The people along the Lena must remember very selective information regarding certain magnitudes of cold to conduct subsistence gathering of fish and water in early winter. Some people even recognize these activities to occur in a specific 2-week period from the end of October to first week of November. This type of knowledge can help residents understand the extreme scale of spring floods through the accumulation of local empirical observations and execution of plans based on these data.

3.6 Conclusions

This paper has presented a detailed picture of the relationship between human inhabitants of a specific area along the Lena River and the river itself. The distinctive feature of this relationship is coexistence between the state of adjustment in the human use system and the state of nature in the natural events system.

The use of certain magnitudes of cold during the freezing process and the expectation of mild spring flooding related to pasture management during the thawing process are the key socio-cultural contexts underlying what has become more frequent extreme spring flooding along the Lena. This is a unique cultural ecology of the local population in the middle Lena River basin.

Severe flooding can have disastrous results such as damaged homes, buildings, roads, and basic infrastructure. Reports from the republic government mainly focus on this type of socioeconomic damage, and primarily concern which preventative policies can be formulated to mitigate future flood damage. However, one must consider the full context of the human–river relationship to understand the background of disastrous flooding. Policymakers should strive to thoroughly comprehend the sociocultural context of the livelihoods of the Lena riverside population to properly decide upon policies to adapt to this flooding.

The indigenous knowledge and sociocultural contexts of the population should be reevaluated in terms of regional disaster management. Currently, that knowledge can only provide residents with the ability to understand potential scenarios inherent in the series of events during river thaw. Unfortunately, the knowledge cannot guide people to take concrete action against a specific disaster. However, the ability to predict feasible scenarios is a simple and important adaptive ability for these individuals. If a number of residents have knowledge that enables them to respond to disastrous events, this in combination with adequate emergency measures provided by local agencies should enhance social resilience against flooding.

My suggestion is that indigenous knowledge regarding the freezing and thawing process should be re-explained with methods using a quantitative range of meteorological conditions that local people can observe. Local concepts such as *kyd'ymamakh* (frazil ice) and *yrbyy* (the appearance of a stream in spring between the riverbank and ice cover) should be used with time-location, quantitative observation. A combination of indigenous knowledge and local meteorology should be implemented. This is necessary to allow the local residents to adapt their indigenous knowledge to the presently changing conditions. The more the residents are able to enhance their ability to predict specific events from their own perspectives, the more latitude these people will have to prepare for a disaster. There must be support to establish legislative and economic policies to prevent damage from the Lena spring flood, from both local and national administrations.

References

Beltaos S, Prowse T, Bonsal B, MacKay R, Romalo L, Pietroniro A, Toth B (2006) Climatic effects on ice-jam flooding of the Peace-Athabasca Delta. Hydrol Process 20:4031–4050

Burakov DA, Avdeeva YV, Kosmakova VF (2007) Floods in Siberian river basins. In: Vasiliev OF et al (eds) Extreme hydrological events: new concepts for security. Springer, New York, pp 111–124

Crate S (2006) Cows, kin and globalization: an ethnography of sustainability. Alta Mira Press, Walnut Creek

Crate S (2008) Gone the bull of winter: grappling with the cultural implications of and anthropology's role(s) in global climate change. Curr Anthropol 49(4):569–595

Filippova VV (2010) Toward the issue on the flood in rivers in Yakutia (in Russian). In: Gumanitarnye nauki v Iakutii: Issledovaniia molodykh uchenykh. Nauka, Novosibirsk, pp 263–270

Hoffman SM, Oliver-Smith A (ed) (2002) Catastrophe and culture: the Anthropology of Disaster. Japanese Edition. (2006). Saigai no jinruigaku (trans: Wakabayashi Yoshihumi). Akashisyoten, Tokyo

Jordan B, Jordan-Bychkov TG (2001) Siberian village: land and life in the Sakha Republic. University of Minnesota Press, Minneapolis
Kilmjaninov V (2007) Hydrological conditions for actions on prevention of ice flooding on the Lena River. In: Vasiliev OF et al (eds) Extreme hydrological events: new concepts for security. Springer, New York, pp 279–284
Kustatov KI, Ammosov AP, Kornilova ZG, Shpakova RN (2012) Anthropogenic factor of ice jamming and spring breakup flooding on the Lena River. Russ Meteorol Hydrol 37(6):392–396
Okladnikov AP (1955) Iakutiia do prisoedineniia k russkomu gosudarstvu. Istoriia Iakutskoi ASSR tom 1. Nauka, Moscow-Leningrad
Oliver-Smith A (2009) Anthropology and the political economy of disasters. In: Jones EC, Murphy AD (eds) The political economy of hazards and disasters. Altamira Press, Lanham, pp 11–28
Prowse TD (2007) River and lake ice. In: United Nations Environmental Programe (ed), Global outlook for Ice and snow. http://www.unep.org/geo/geo_ice. Accessed 16 Oct 2015
PRS – Pravitel'stvo Respublika Sakha (Iakutiia) (2010) Postanovlenie to 27 Maia 2010g. No.253 "Ob utverzhdenii kontsptsii zashchity nasselennykh punktov i ob"ektov ekonomicheski Respublika Sakha (Iakutiia) ot navodnenii i drugikh vidov negativnogo vozdeistviia vod"
Rakkolainen M, Tennberg M (2012) Adaptation in Russian climate governance. In: Tennberg M (ed) Governing the uncertain: adaptation and climate in Russia and Finland. Springer, New York, pp 39–54
Reginony (2004) Reginony Rossii. Osnovnye kharakteristiki sub'ektov Rossiiskoi Federatsii. Ofitsial'noe izdanie. 2004. Stat. Sb. Rosstat, Moscow
Reginony (2010) Reginony Rossii. Osnovnye kharakteristiki sub'ektov Rossiiskoi Federatsii. Ofitsial'noe izdanie. 2010. Stat. Sb. Rosstat, Moscow
Reycraft R (2000) Introduction: environmental disaster and the archaeology of human response. In: Bawdena G, Reycraft R (eds) Environmental disaster and the archaeology of human response. Maxwell Museum of Anthropology, Albuquergue, pp 1–10
Rozhdestvenskii AV, Buzin VA, Shalashina TL (2008) Forming conditions and probable values of maximum water levels of the Lena River near Yakutsk. Russ Meteorol Hydrol 35(1):54–61
Saito S (1985) Tundora to Taiga no sekai: Shiberia no shizen to genshi bunka [Tundra and Taiga: Nature and Archaic culture of Siberia]. Chijinsyobo, Kyoto
Stammler-Gossmann A (2012) The big water of a small river: flood experiences and a community agenda for change. In: Tennberg M (ed) Governing the uncertain: adaptation and climate in Russia and Finland. Springer, New York, pp 55–82
Steward J (1955) Theory of culture change: the methodology of multilinear evolution. University of Illinois Press, Urbana
Sukhoborov V (2006) Problems linked to the prevention of natural disasters in the Sakha Republic (Yakutia). In the conference of "The 3rd northern forum to the prevention of natural disasters in the Sakha Republic (Yakutia) 2006". http://www.yakutiatoday.com/events/inter_FWG_emercom.html. Accessed 23 Aug 2012
Takakura H (2009) The concept of manhood in post-socialist Siberia: the Sakha father as a wise hunter and a pastoralist. Sibirica 8-1:45–67
Takakura H (2010) The social and symbolic construction of alaas landscapes in the Siberian Forest among the Sakha. Acta Slavica Iaponica 28:51–70
Takakura H (2015) Arctic Pastoralist Sakha: the ethnography of evolution and micro-adaptation in Siberia. Trans Pacific Press, Melbourne
Tanaka S, Hayashi H (1998) Toward the anthropology of disaster: a trial for disaster ethnography and its systemization [in Japanese]. Chiiki anzen gakkai ronbun hokokusyu [Papers of the Annual Conference of the Institute of Social Safety Science] 8:14–19
White G (1974) Natural hazards research: concepts, methods, and policy implications. In: White G (ed) Natural hazards: local, national, global. Oxford University Press, Oxford, pp 3–16
Yoshikawa Y, Watanabe K, Hayakawa H, Hirai K (2011) Field observation of a river ice jam in the Shokotsu River (in Japanese). Kanchi doboku kenkyujo geppo [Monthly reports of Civil engineering research institute for cold region] 701:2–9

Chapter 4
Monitoring Spring Floods on the Lena River Using Multiple Satellite Sensors

Toru Sakai, Hiroki Takakura, Makoto Okumura, Sigemi Hatta, Yasuhiro Yosikawa, Tetsuya Hiyama, and Yasushi Yamaguchi

Abstract Riparian regions in Siberia have experienced extreme floods in the last decade because winter precipitation has increased. However, detailed statistical information about spring floods is lacking. Remote sensing is an ideal tool for collecting data over large areas. The objective is to evaluate the relative usefulness and limitations of multiple satellite sensors for monitoring spring flooding across the entire Lena River. Spring floods progress at a speed of ~100 km day^{-1} during the snowmelt period, expanding the width of the floodplain from several kilometers upstream to several tens of kilometers downstream. The 30-m-resolution Landsat and 500-m-resolution MODIS were sufficient to monitor the spatial extent of the flood. However, the upstream floodplain was too narrow to detect using 25-km-resolution AMSR-E. The AMSR-E was barely able to detect the presence or absence of floods at midstream. However, the AMSR-E microwave sensor could monitor day-to-day variation of spring floods. Images from optical sensors such as Landsat and MODIS were considerably limited by cloud cover. The 10-m-resolution

T. Sakai (✉)
National Institute for Environmental Studies, 2-16 Onogawa, Tsukuba 305-8506, Japan
e-mail: sakai.toru@nies.go.jp

H. Takakura
Center for Northeast Asian Studies, Tohoku University,
Kawauchi 41, Aobaku, Sendai 980-8576, Japan

M. Okumura
International Research Institute of Disaster Science,
Tohoku University, Sendai 980-8579, Japan

S. Hatta
Tomakomai National College of Technology, Tomakomai 059-1275, Japan

Y. Yosikawa
Kitami Institute of Technology, Kitami 090-8507, Japan

T. Hiyama
Institute for Space-Earth Environmental Research,
Nagoya University, Nagoya 464-8601, Aichi, Japan

Y. Yamaguchi
Graduate School of Environmental Studies, Nagoya University, Nagoya 464-8601, Japan

T. Hiyama, H. Takakura (eds.), *Global Warming and Human - Nature Dimension in Northern Eurasia*, Global Environmental Studies,
DOI 10.1007/978-981-10-4648-3_4

PALSAR was unable to monitor spring floods during the 5-year operational period, because the temporal resolution of 46 days was insufficient to monitor floods. It was sometimes possible to obtain meaningful results from a single remote sensor, but it was impossible to fully understand spring flood behavior over the entire Lena River. An assemblage of sensors with different spatial, temporal, and spectral resolutions would be helpful for flood risk management.

Keywords Flood • Global warming • Lena River • Remote sensing • River ice • Snowmelt

4.1 Introduction

The high latitudes of the Northern Hemisphere are among the regions most vulnerable to climate change. Annual mean air temperatures have risen by 2–4 °C over the last century. Such an increase in temperature may have a large impact on the water balance by influencing evapotranspiration. Future predictions suggest that precipitation will also increase because of the greater amount of water vapor held in a warmer atmosphere, but this increase may not occur in all regions (IPCC 2013). The increase in precipitation has already been observed in Siberia (Iijima et al. 2014; Vey et al. 2013). In the context of future climate–biosphere interactions, a number of studies have reported that the increased precipitation could have negative impacts, such as biodiversity reduction and permafrost degradation in Siberia (Fedorov et al. 2014; Iijima et al. 2014; Ohta et al. 2014). Therefore, significant concerns have been raised about the change in precipitation.

Increased precipitation affects not only various ecosystem components but also human activities through extreme hydrologic events. In particular, an increase in winter precipitation enhances the probability of spring floods in Arctic regions (Beltaos 2007, 2012). Winter precipitation affects the distribution and dynamics of snow, river and lake ice, and permafrost through associated unique radiative and thermal properties. Although the water volume of winter precipitation is less than that of summer precipitation, snow coverage has a large spatial extent in the Northern Hemisphere. Therefore, snowmelt can be a source of extreme floods (Prowse and Beltaos 2002). In fact, riparian regions have experienced extreme floods in spring during the last decade. The consequences of flooding include multiple types of losses, such as environmental and economic losses and loss of life. The economic losses associated with flood in 2001 amounted to ~21% of the revenue of the Sakha Republic in the Russian Federation (Takakura 2012). Assessment and management of flood risk related to climate change are important for ensuring economic and environmental quality, human wellbeing, and security. Several studies have therefore developed flood forecasting and warning systems, using parameters based on field measurements such as hydrographs, ice thickness and strength, snow depth, and cumulative temperature in degree-days (Ma and Fukushima 2002; Prowse and

Conly 1998; Shulyakovskii 1972). However, hydrologic responses to climate change vary significantly with basin size, location, and topography. Although plot-scale field measurements are valuable, that approach is costly, spatially limited, and difficult to extrapolate over an entire river basin, especially in Arctic regions.

Remote sensing is considered an ideal tool for collecting data toward understanding ecological and environmental processes over a range of scales. Methodologies for detecting flood spatial and temporal distributions have been developed using various types of satellite sensors (Arnesen et al. 2013; Khan et al. 2011; Pavelsky and Smith 2004; Sakai et al. 2015; Sakamoto et al. 2007; Xiao et al. 2002). However, there is a tradeoff among different resolutions (i.e., in terms of spatial, temporal, and spectral details). Such resolution differences affect the accuracy of land-cover classification and subsequent interpretation (Cihlar 2000; Kerr and Ostrovsky 2003; Lu and Weng 2007). For example, the use of time series data from wide field-of-view sensors such as the Moderate Resolution Imaging Spectroradiometer (MODIS) provides near-daily flood coverage over an entire river basin (Pavelsky and Smith 2004). Datasets updated with high temporal frequency are important for understanding the flood dynamics, but coarse spatial-resolution sensors often make it difficult or impossible to map the spatial extent of floods in narrow riparian areas. Accurate sub-pixel information is obtained from medium- to high-spatial-resolution sensors such as Landsat TM/ETM+ (Sakai et al. 2015), but the images derived from such sensors are not updated frequently because of long revisit periods. Data availability and quality (e.g., as affected by cloud cover) associated with image acquisition at optimal times are also factors. If the time interval between images covering the same geographic area is too long, it is possible that some phenomena will be missed. Microwave sensors such as Phased Array type L-band Synthetic Aperture Radar (PALSAR) and Advanced Microwave Scanning Radiometer – EOS (AMSR-E) have several advantages in persistently cloudy areas (e.g., the humid tropics or high latitudes), owing to the ability to operate day and night through cloud cover (Arnesen et al. 2013; Khan et al. 2011).

Remote sensing provides a wide range of information over large areas and long time periods, and there is little doubt that it is a potentially powerful tool for assessment and management of flood risk. However, the development of flood monitoring is challenging, because it requires images with wide geographic coverage, adequate spatial resolution, high temporal resolution, and minimal cost. Specific issues pertaining to various types of sensors have not yet been explored systematically. The objective of this chapter is to evaluate the relative utility and limitations of multiple satellite sensors (MODIS, Landsat TM/ETM+, AMSR-E, and PALSAR) for monitoring spring flooding over the entire Lena River basin. Success of flood risk management relies heavily on an understanding of flood behavior in response to extreme climate events. This study is one of many steps toward our long-term goal of developing sustainable flood risk management strategies against climate change.

4.2 Study Area

With a length of 4400 km, the Lena River is the 10th longest river in the world (Fig. 4.1). From its source in the mountains along the western bank of Lake Baikal (latitude ~53°N), the Lena flows northeastward before changing direction to flow northward in a large arc around Yakutsk, the capital of the Sakha Republic of the Russian Federation (62°02′N, 129°43′E). It then flows to its delta in the Arctic Ocean via the Laptev Sea. The Lena River is navigable from near its source to its mouth, without any dams or human construction. It discharges 12 million tons of sediment and 540 km^3 of water annually. Its flow plays a dominant role in the Arctic hydrologic cycle and oceanic freshwater budget (Aagaard and Carmack 1989). The climate of the river basin is continental and is characterized by very low air temperatures and little precipitation. Annual mean air temperature is below the freezing point, although the temperature in summer is moderately warm (Fig. 4.2). River ice can thicken to more than 100 cm. Major vegetation types between 55°N and 65°N include tundra in the north and taiga in mountainous areas, with steppe, cropland, and temperate forest in the south.

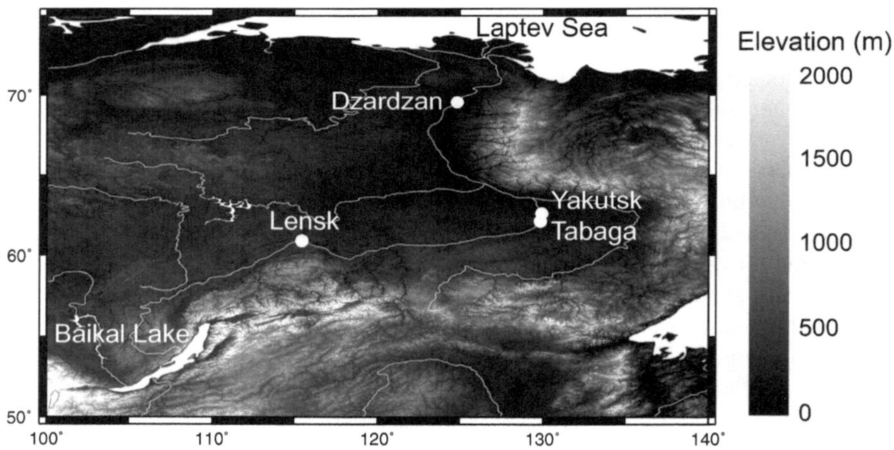

Fig. 4.1 Global 30-arc-second elevation map (GTOPO30) of Lena River area

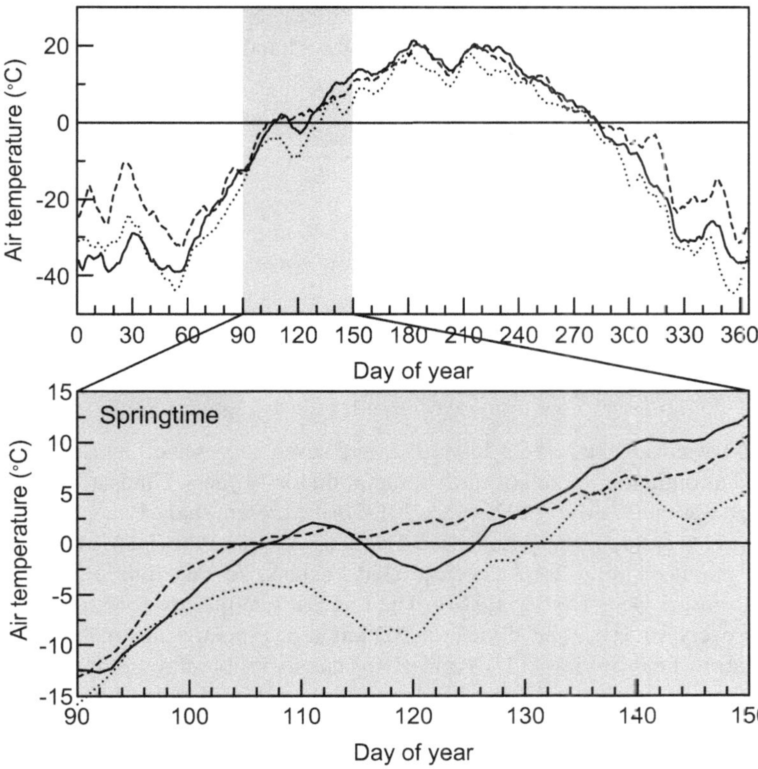

Fig. 4.2 Ten-day moving average of air temperature at upstream (Lensk, *dashed line*), midstream (Yakutsk, *solid line*), and downstream (Dzardzan, *dotted line*) meteorological stations on Lena River in 2007. Gray color shows springtime (April–May; day-of-year (DOY) 090–150)

4.3 Sensor Overview

4.3.1 MODIS

The MODIS sensor onboard the Terra and Aqua satellites has observed the earth surface since 18 December 1999. MODIS has 36 spectral bands, with detectors fixed on four focal-plane assemblies. MODIS bands 1–19 and 26, covering spectral wavelengths from 412 to 2100 nm, are reflective solar bands that produce images from daytime-reflected solar radiation. Bands 20–25 and 27–36, with spectral coverage 3700 to 14,400 nm, are thermal emissive bands that facilitate daytime and nighttime observations of the earth's thermal emissions. Spatial resolutions of MODIS at nadir are 250 m for bands 1 and 2 (40 detectors per band), 500 m for bands 3–7 (20 detectors for each), and 1 km for bands 8–36 (10 detectors per band). MODIS spectral bands are aligned in the cross-track direction on the focal-plane assemblies, and the detectors are aligned in the along-track direction. Two types of products were used herein, daily (MOD09GA) and 8-day composite (MOD09A1) surface reflectance.

We used the Normalized Difference Snow Index (NDSI) to monitor the extent of snow cover using a visible band such as MODIS band 4 and a short-wave infrared band such as MODIS band 6:

$$NDSI = \frac{B4 - B6}{B4 + B6}, \tag{4.1}$$

where B4 and B6 are reflectances at 555 and 1640 nm, respectively. The use of band 6 allows effective discrimination between snow cover and clouds.

4.3.2 Landsat TM/ETM+

Landsat-1 was launched on 23 July 1972, and seven successive Landsat series have provided a continuous record of earth observation for 40 years. Landsat-8, the latest in the series, was launched on 11 February 2013. In the present study, Landsat-5 Thematic Mapper (TM) and Landsat-7 Enhanced Thematic Mapper Plus (ETM+) images were used to monitor spring 2007 floods around Yakutsk. At that time, the operational Landsat sensors were TM and ETM+. The Landsat satellites overpass every location on earth every 16 days at an altitude of 705 km in near-polar, sun-synchronous orbits with a narrow field-of-view (15°). Landsat TM has six visible-through-middle-infrared bands with a 30-m resolution and one thermal-infrared band with a 120-m resolution. Landsat ETM+ has multispectral bands similar to TM but with a 60-m resolution thermal-infrared band and a new 15-m resolution panchromatic band. Digital numbers in six spectral bands from visible to middle-infrared were converted into surface reflectance to normalize solar illumination and viewing geometry. The earth–sun distance was used to adjust for seasonal variations in solar irradiance.

4.3.3 AMSR-E

AMSR-E, operated by the Japan Aerospace Exploration Agency (JAXA), is a passive microwave radiometer mounted on the Aqua satellite. It collected data since a launch on 4 May 2002, but a fault halted measurements on 4 October 2011. The Global Change Observation Mission - Water (GCOM-W) satellite launched on 18 May 2012 continues the legacy of AMSR-E, with AMSR2. The AMSR-E is an eight-frequency (6.9, 10.6, 18.7, 23.8, 36.5, 50.3, 52.8, and 89.0 GHz) total-power microwave radiometer with dual polarization (except for two vertical channels in the 50-GHz band) and daily ascending (13:30 equatorial local crossing time) and descending (01:30 equatorial local crossing time) overpasses. Swath width is 1445 km and grid resolution is 25 km.

We developed flood inundation maps from the simplest algorithm, the index of surface wetness (ISW), derived using AMSR-E L3 data from the ascending path:

$$ISW = \frac{T_{36.5H} - T_{6.9H}}{T_{6.9H}} \quad (4.2)$$

where $T_{6.9H}$ and $T_{36.5H}$ are horizontal brightness temperatures at 6.9 and 36.5 GHz, respectively (Koike et al. 1996).

4.3.4 *ALOS PALSAR*

JAXA launched the Advanced Land Observing Satellite (ALOS) on 24 January 2006, ensuring a 46-day repeat cycle. PALSAR is one of the instruments onboard ALOS and is an enhanced version of the Japanese Earth Resources Satellite (JERS-1) SAR instrument. ALOS PALSAR-1 was retired on 12 May 2011, and PALSAR-2 was the follow-on in 2014. PALSAR has a frequency of 1270 MHz (23.6 cm, i.e., L-band) and chirp bandwidths 14 and 28 MHz. The instrument operates in five observation modes: (i) fine-beam single (FBS) and (ii) fine-beam dual polarization (FBD); (iii) polarimetric (PLR); (iv) ScanSAR; (v) direct transmission (DT). However, because of the observation schedule of ALOS PALSAR-1, data in the FBS and FBD modes were available only in early spring for Siberia. Slant range resolutions of the FBS and FBD modes are 4.7 and 9.4 m, respectively. Azimuthal resolution irrespective of mode is 3.1 m, with an incidence angle of 39°.

4.4 Results and Discussion

4.4.1 *Timing of Snowmelt and Spring Flood*

MODIS and AMSR-E achieve near-daily coverage over the global surface, thereby providing an opportunity for global monitoring of terrestrial ecosystems (Fig. 4.3). However, one major limitation of optical sensors such as MODIS is an inability to penetrate clouds (Fig. 4.3a), but the longer wavelength pulses of AMSR-E can do so (Fig. 4.3b). The land surface was obscured by clouds on most days, especially in polar or equatorial regions of Russia, Southeast Asia, Africa, and South America. A number of studies have reported difficulty in obtaining cloud-free area-wide coverage every day in cloud-prone areas (Goward et al. 2006; Pesaresi 2000; Ju and Roy 2008; Marshall et al. 1994). A temporal composite product using high-frequency data provides an image that is almost cloud-free. Figure 4.4 shows 8-day composite images of MODIS NDSI for the Lena River basin during springtime, from April to May 2007. Values of NDSI > 0.4 typically indicate the presence of snow (Hall et al. 2002). Therefore, white areas in Fig. 4.4 are nearly all snow, not cloud. The entire basin was covered by snow in early April (Fig. 4.4a–c), and the Lena River was frozen because air temperature was below freezing (Fig. 4.2). Snowmelt began upon reaching 0 °C during late April in upstream areas (Figs. 4.2 and 4.4d). Snow coverage then decreased with increasing air temperature, and the rate of decrease

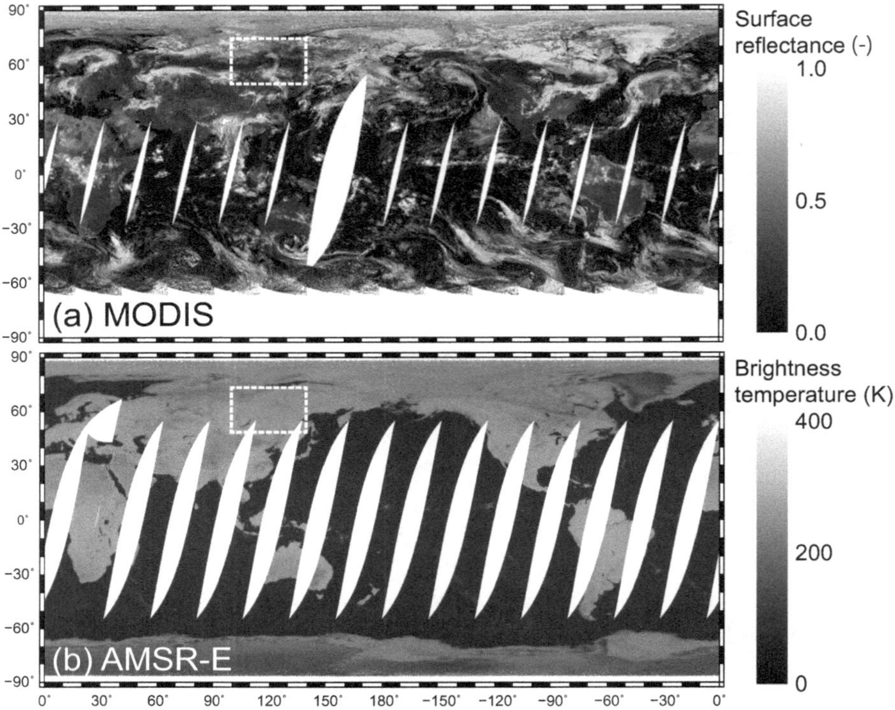

Fig. 4.3 Global maps of (**a**) MODIS surface reflectance (band 1, 620–670 nm) and (**b**) AMSR-E brightness temperature (6.9 GHz-H) on 15 May 2007 (DOY 135). *Rectangle* shows study area

accelerated in early May (Fig. 4.4e, f). Much snow cover disappeared upstream, with the exception of mountain regions at latitudes 55–60°N, although almost all snow cover persisted downstream (Figs. 4.1 and 4.4f). Downstream snow cover disappeared completely in June (Fig. 4.4g, h). Regional air temperature showed a distinct south–north gradient (Fig. 4.2). Annual air temperatures in 2007 at upstream (Lensk), midstream (Yakutsk), and downstream (Dzardzan) meteorological stations were −3.4, −7.1, and −10.2 °C, respectively. Temperatures were colder at higher latitudes. Therefore, snowmelt proceeded in a northerly direction from upstream to downstream. The date when air temperature consistently exceeded 0 °C differed by as much as 30 days between upstream and downstream (Fig. 4.2). The Lena River was frozen for 6–7 months of the year in the south and 7–8 months in the north.

Snowmelt is related to river discharge (Rawlins et al. 2005). A large amount of upstream snowmelt enhances the probability of extreme floods, because river ice is still frozen hard downstream and ice blocks river flow (Fig. 4.4). Thus, the magnitude of spring flooding is greatly changed by dates and locations of snowmelt (Beltaos 2007; Beltaos 2012; Prowse and Beltaos 2002). Figure 4.5 shows temporal variation in water level at Tabaga in 2007. Water was very stable at a low level of 1 m when river ice was still frozen but the level suddenly increased to 10.6 m upon

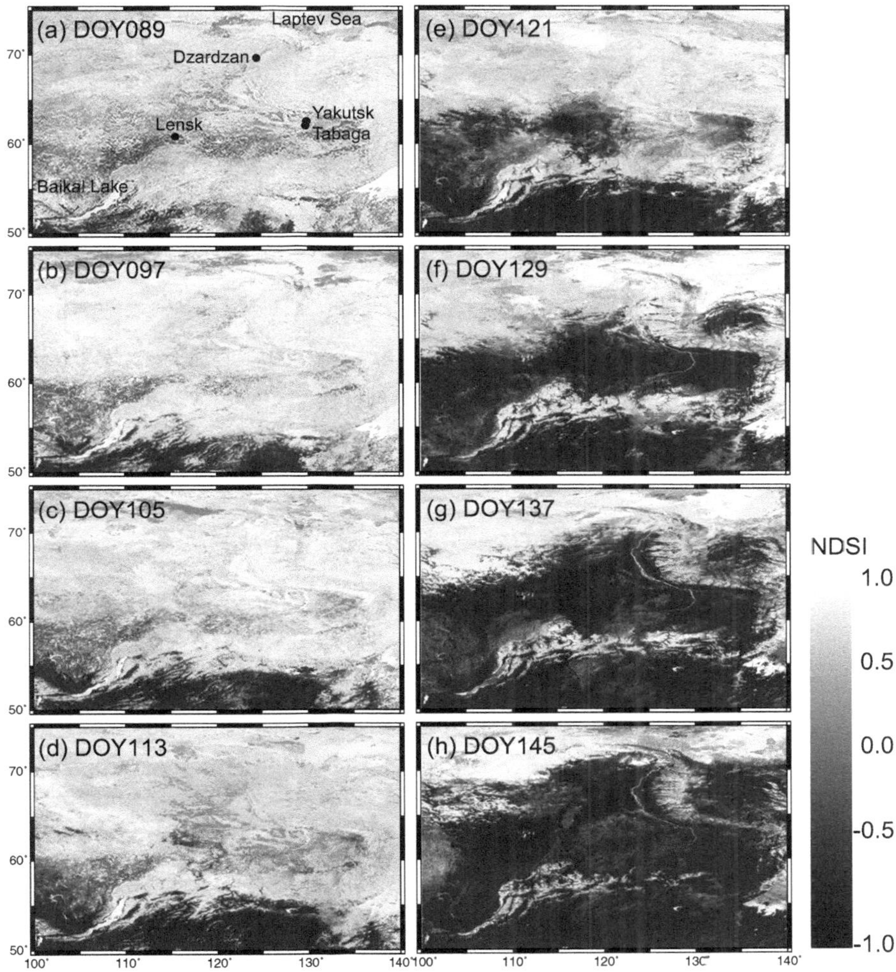

Fig. 4.4 Eight-day composite MODIS NDSI images of Lena River area during 2007 spring (April–May) (DOY 089–145)

river ice breakup. A flood occurs when the water level exceeds a predefined level. At Tabaga, the danger level was approximately 8 m. A second flood occurred in June because of snowmelt in the mountain areas (Figs. 4.1, 4.4 and 4.5). Remote sensing cannot directly detect the amount of snowmelt. However, it can map the spatial extent of snow cover and subsequent spring floods, reflecting variations in climate and hydrology. Because Fig. 4.4 is based on images of the 8-day composite at a spatial resolution of 500 m, it is not possible to represent the exact date and location of flood occurrence. However, these images can identify the starting and ending weeks of flooding throughout the Lena River basin.

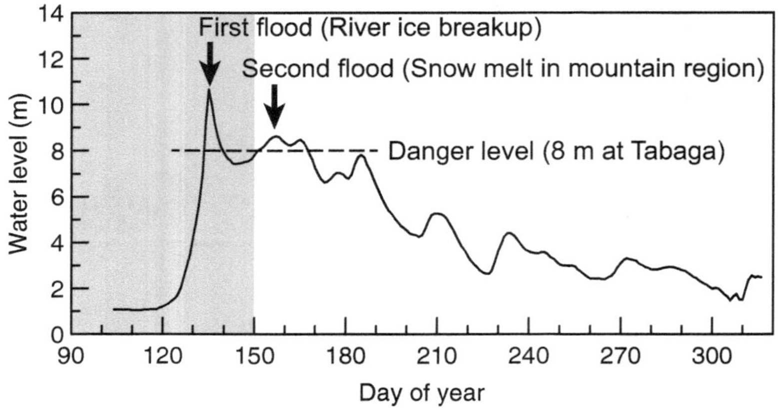

Fig. 4.5 Temporal variation in water level at Tabaga in 2007. *Gray color* shows springtime (April–May) (DOY090–150)

4.4.2 Spatial Extent of Spring Flood

Spring floods occurred at various locations along the Lena River during spring. Around Yakutsk, there was a spring flood during 13–15 May 2007 (day of year [DOY] 133–135). Multiscale images were collected to compare differences in images of the flood. Figure 4.6 shows the spatial extent of the flood using Landsat and MODIS within a 25-km pixel of AMSR-E. Unfortunately, PALSAR images at the same time could not be acquired because of the long revisit period (46 days). Fortunately, however, three consecutive Landsat images were acquired using the overlap area between neighboring path images of the TM and ETM+, despite the Landsat revisit period of 16 days. There was a high contrast in spectral signature between river ice and open water. The former appeared white and the latter black in the gray-scale, false-color images (Fig. 4.6). On 13 May 2007 (DOY 133), the Lena River was mostly covered by sheet ice, and open water appeared only in small patches (Fig. 4.6a). Ice breakup was observed at the junction of the main river and a tributary, with ice in that tributary breaking up earlier than in the main stream. Landsat with a spatial resolution of 30 m could map ice conditions of the tributary, but MODIS with a spatial resolution of 500 m could not (Fig. 4.6b). AMSR-E with a spatial resolution of 25 km could not map the 1-km wide mainstem. On 14 May 2007 (DOY 134), there was a major flood around Yakutsk. The Lena River was covered mainly by rubble ice, and open water spread onto floodplains. The river spread across a wide area, and its width increased by as much as 15 km. Landsat had sufficient spatial resolution not only to map the floodplain distribution but also to monitor the shape, size, and location of relatively large rubble ice (Fig. 4.6c). However, it was difficult to distinguish rubble from sheet ice using MODIS (Fig. 4.6d). The wide floodplain around Yakutsk could barely be detected using a single AMSR-E pixel. The brightness temperature at 36.5 GHz, which is highly

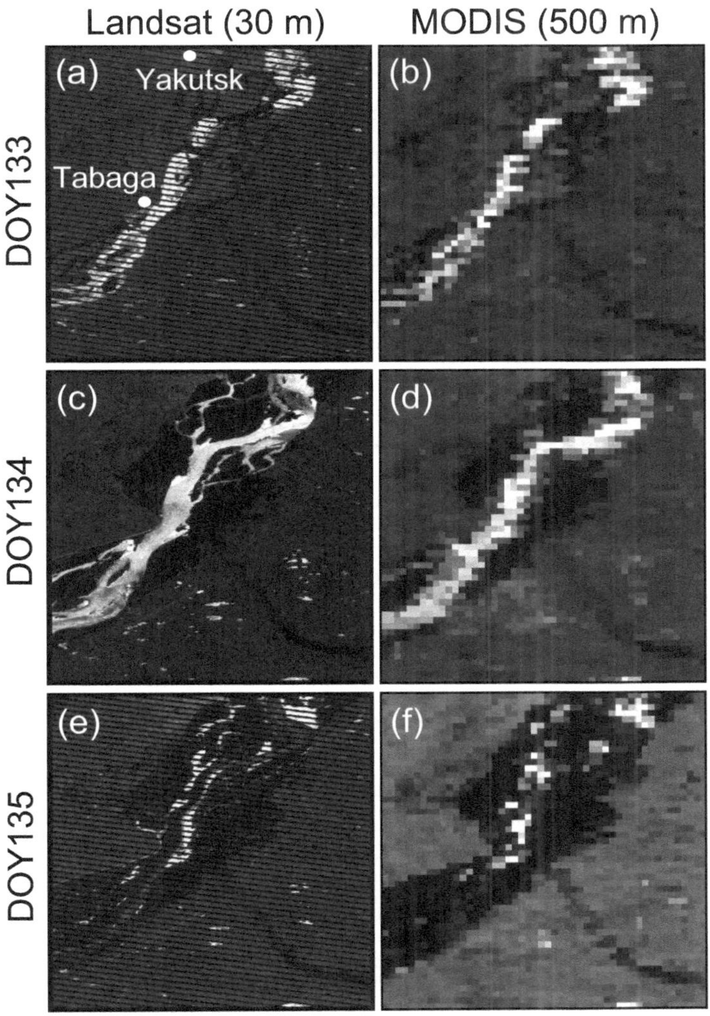

Fig. 4.6 Spatial extent of spring flood as depicted by 30-m Landsat and 500-m MODIS within a 25-km pixel of AMSR-E, 13–15 May 2007 (DOY 133–135). Gray-scale, false-color images are displayed

sensitive to moisture content, decreased from 263 to 259 K. On 15 May 2007 (DOY 135), almost all rubble ice was swept downstream, the river was covered by open water, and the floodplain was slightly larger than on the previous day (Fig. 4.6e, f).

Visual comparison among multiple satellite images revealed a variety of features at various spatial resolutions (Fig. 4.6). Spatial resolution was important for detecting spring flooding in narrow riparian areas. More accurate flood maps could be extracted effectively from higher spatial-resolution images. The 30-m resolution of

Landsat was sufficient to monitor the spatial extent of the spring flood on the Lena. The 500-m MODIS images also revealed relatively clear flood maps, but the 25-km AMSR-E images were composed of a mixture of land surface types.

Higher spatial-resolution images have the potential to extend flood risk management and land-use planning to a local scale. However, the availability and utility of the images must be considered. For example, the coverage area varies with the spatial resolution of the image (Franklin and Wulder 2002). Seventy-five Landsat-like images (e.g., with a 30-m spatial resolution and a 180-km-wide swath) or 494 PALSAR-like images (e.g., with a 10-m spatial resolution and a 70-km-wide swath) is needed to cover the entire Lena Basin (2,420,000 km^2), assuming no overlap between neighboring path images. In reality, some overlap areas exist, increasing the total number of images required. Higher spatial-resolution sensors may not be feasible for monitoring rapid changes in spring flooding owing to infrequent temporal coverage. In fact, PALSAR never monitored a spring flood on the Lena during 5 years of operation, because its observation schedule was designed to be in fine mode during spring. If it was designed for ScanSAR mode, there might have been some possibility of acquiring PALSAR images at the time of the flood. The ability to use multiple images acquired over a single region is important for characterizing spatial patterns, detecting changes, and relating patterns to processes (Wu and Hobbs 2002).

4.4.3 Temporal Variations of Spring Flood

Monitoring the temporal variation of spring floods requires sensors with high temporal resolution. MODIS and AMSR-E could cover the entire Lena River every day because of their wide swath widths (Fig. 4.3). Figure 4.7 shows temporal changes in the spring flood using AMSR-E. In the AMSR-E images, inundation area along the river was indicated by exceedance of an ISW threshold of 0.07, and such inundation areas during floods were overlaid (Fig. 4.7). The microwave sensor could readily discriminate the land surface every day and night, even under cloud cover. However, the spring flood could not be detected in upstream areas through the narrow valley, although the flood had a scale of several kilometers. Upstream floods were too small relative to the size of an AMSR-E pixel. The spring flood could be detected in the midstream area around Yakutsk on 14 May 2007 (DOY 134), because the floodplain extended more than 15 km (Fig. 4.6). After that date, the flood was monitored daily to the north. The flood arrived at the river mouth and began entering the Laptev Sea on 31 May 2007 (DOY 151). The average progression speed was estimated at ~100 km day^{-1}, based on the relationship between distance traveled and number of days. However, the progression speed was affected by river ice conditions. The Lena River spread over a very wide area, where the flow was interrupted by ice jams. River width increased up to 50 km downstream. Therefore, it was possible to attain meaningful results for understanding flood behavior using coarse spatial resolution images such as AMSR-E.

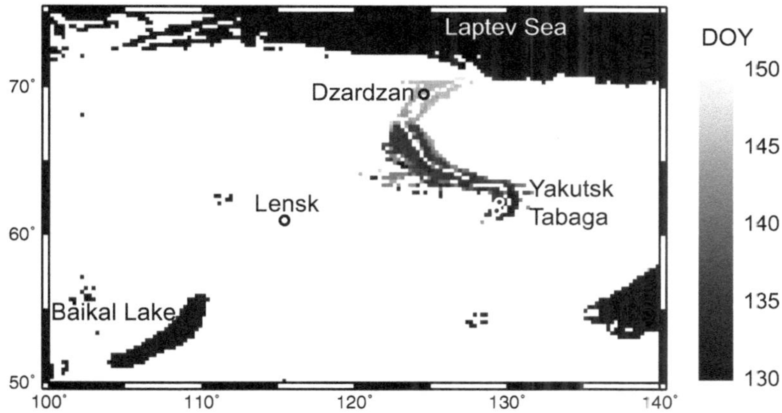

Fig. 4.7 Detection date (in DOY) of 2007 spring flood on Lena River using AMSR-E

Ice jams often cause serious floods because they drastically increase the water level at their locations for a few minutes or many days. The release of ice jams is also a concern, because a large volume of stored water can be released at once (Beltaos 2007, 2012; Prowse and Beltaos 2002). Two-thirds of all flood damage is attributable to ice jams (Humes and Dublin 1988). An ice-jam flood was monitored around Yakutsk during mid-May 2011. Nearly cloud-free MODIS images were acquired for 10 days, although the effects of cloud cover persisted in some areas (Fig. 4.8). An ice jam was observed at a bend of the Lena River on 13 May 2011 (DOY 133; Fig. 4.9). The next day, there was severe flooding around Yakutsk (Fig. 4.8d). The length of the ice jam was > 100 km, although it shrank over time (Fig. 4.8d–g). The ice jam blocked river flow until 17 May 2011 (DOY 137). Therefore, many villages not protected by dikes were impacted by widespread ice-jam flooding over several days (Fig. 4.10). The inundated water began receding on 19 May 2011 (DOY 139; Fig. 4.8i).

Daily flood monitoring is valuable for understanding dynamic processes that determine river behavior. Flood management operations would be enhanced by greater temporal availability of coarse spatial-resolution sensors. Such sensors with the minimum map unit required to detect floods would also assist in defining acquisition strategies for higher spatial-resolution images.

4.5 Conclusions

The use of remote sensing to monitor landscape change has become widespread. In the present study, we evaluated the relative utility and limitations of monitoring spatial and temporal patterns of spring floods over the entire Lena River basin using multiple remote sensing sensors.

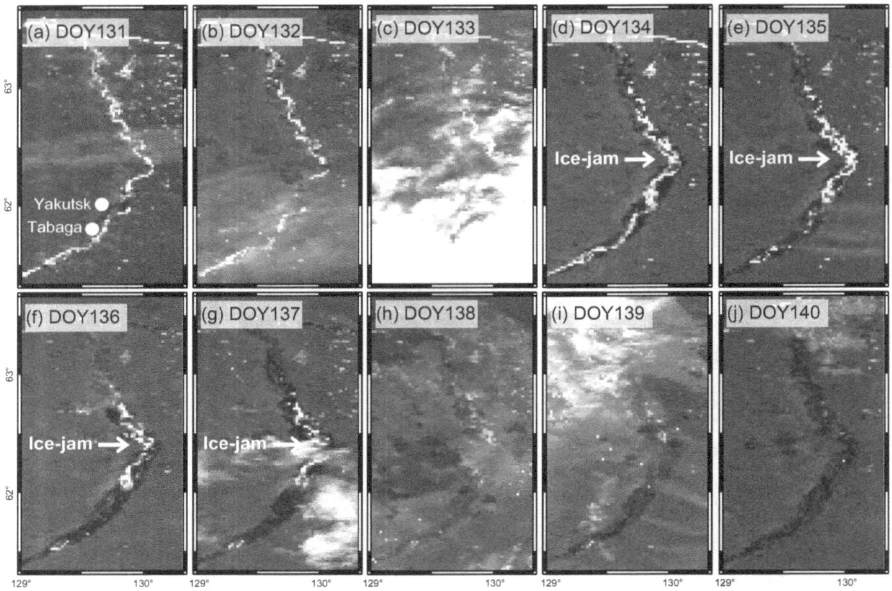

Fig. 4.8 Monitoring of ice-jam flood around Yakutsk during 11–20 May 2011 (DOY 131–140) using MODIS. *Gray-scale, false-color* images are shown (R:G:B = band2:band4:band1)

Fig. 4.9 River ice jam around Yakutsk on 12 May 2011 (DOY 133)

Fig. 4.10 Long-term flood damage by ice jam around Yakutsk on 17 May 2011 (DOY 137)

Spring floods occurred during the snowmelt period from late April through early June. Snowmelt was earlier in the upstream portion of the river, because of a north–south difference in air temperature. That earlier upstream snowmelt triggered a spring flood, because downstream river ice was still frozen and blocked river flow. The flood progressed from upstream to downstream, expanding the floodplain from several kilometers upstream to several tens of kilometers downstream. The 30-m-resolution of Landsat was sufficiently small to detect the spatial extent of the flood. However, the 10-m-resolution PALSAR never monitored a spring flood along the Lena during 5 years of operation. This is because its temporal resolution of 46 days was insufficient to detect the flood across the entire river over approximately 1 or 2 months. High spatial-resolution sensors could not resolve rapid changes in the temporal dynamics of the flood. In contrast, coarse spatial-resolution sensors such as MODIS and AMSR-E could monitor daily flood dynamics. Day-to-day monitoring of flooding is valuable for understanding the dynamic processes that determine river behavior. In particular, AMSR-E has great potential for mapping and monitoring because of its ability to operate day and night through cloud cover. However, its 25-km resolution barely detected the wide floodplain, and the upstream floodplain was too narrow for detection.

It was sometimes possible to obtain meaningful results from limited available information with the use of a single remote sensing sensor, but it was absolutely impossible to fully understand spring flood behavior over the entire Lena Basin. The greatest benefits are likely to come from selecting appropriate satellite sensors for the intended use rather than from improving flood-detection algorithms. Information from multiple sensors should be combined to achieve reasonable approximation. An assemblage of sensors with various spatial, temporal, and spectral resolutions would help identify the extent of spring floods. The information presented herein provides a visual understanding of the dynamic processes of spring flooding that will assist local policymakers in making sound decisions regarding sustainable flood risk management.

References

Aagaard K, Carmack EC (1989) The role of sea ice and other fresh water in the Arctic circulation. J Geophys Res Oceans 94(C10):14485–14498

Arnesen AS, Silva TSF, Hess LL, Novo EMLM, Rudorff CM, Chapman BD et al (2013) Monitoring flood extent in the lower Amazon River floodplain using ALOS/PALSAR ScanSAR images. Remote Sens Environ 130:51–61

Beltaos S (2007) River ice breakup processes: recent advances and future directions. Can J Civ Eng 34(6):703–716

Beltaos S (2012) Distributed function analysis of ice jam flood frequency. Cold Reg Sci Technol 71:1–10

Cihlar J (2000) Land cover mapping of large areas from satellites: status and research priorities. Int J Remote Sens 21(6–7):1093–1114

Fedorov AN, Gavriliev PP, Konstantinov PY, Hiyama T, Iijima Y, Iwahana G (2014) Estimating the water balance of a thermokarst lake in the middle of the Lena River basin, eastern Siberia. Ecohydrology 7(2):188–196

Franklin SE, Wulder MA (2002) Remote sensing methods in medium spatial resolution satellite data land cover classification of large areas. Prog Phys Geogr 26:173–205

Goward S, Arvidson T, Williams FJ, Irons J, Franks S (2006) Historical record of Landsat global coverage: mission operations, NSLRSDA, and international cooperator stations. Photogramm Eng Remote Sens 72:1155–1169

Hall DK, Riggs GA, Salomonson VV, DiGirolamo NE, Bayr KJ (2002) MODIS snow-cover products. Remote Sens Environ 83(1–2):181–194

Humes TM, Dublin J (1988) A comparison of the 1976 and 1987 Saint-John River Ice Jam flooding with emphasis on antecedent conditions. In: Proceedings of the 5th workshop on river ice/ice jams, pp 43–61

Iijima Y, Ohta T, Kotani A, Fedorov AN, Kodama Y, Maximov TC (2014) Sap flow changes in relation to permafrost degradation under increasing precipitation in an eastern Siberian larch forest. Ecohydrology 7(2):177–187

IPCC (2013) Climate change 2013: The physical science basis. (Contribution of Working Group I to the Fifth Assessment Report of the Intergovernmental Panel on Climate Change). Cambridge University Press, Cambridge

Ju J, Roy DP (2008) The availability of cloud-free Landsat ETM+ data over the conterminous United States and globally. Remote Sens Environ 112(3):1196–1211

Kerr JT, Ostrovsky M (2003) From space to species: ecological applications for remote sensing. Trends Ecol Evol 18(6):299–305

Khan SI, Yang H, Wang J, Yilmaz KK, Gourley JJ, Adler RF et al (2011) Satellite remote sensing and hydrologic modeling for flood inundation mapping in lake victoria basin: implications for hydrologic prediction in ungauged basins. IEEE Trans Geosci Remote Sens 49(1):85–95

Koike T, Tsukamoto T, Kumakura T, Lu M (1996) Surface wetness. In: Proceedings of international workshop on macro-scale hydrological modeling, pp 87–96

Lu D, Weng Q (2007) A survey of image classification methods and techniques for improving classification performance. Int J Remote Sens 28(5):823–870

Ma X, Fukushima Y (2002) A numerical model of the river freezing process and its application to the Lena River. Hydrol Process 16(11):2131–2140

Marshall GJ, Dowdeswell JA, Rees WG (1994) The spatial and temporal effect of cloud cover on the acquisition of high quality landsat imagery in the European Arctic sector. Remote Sens Environ 50(2):149–160

Ohta T, Kotani A, Iijima Y, Maximov TC, Ito S, Hanamura M et al (2014) Effects of waterlogging on water and carbon dioxide fluxes and environmental variables in a Siberian larch forest, 1998–2011. Agric For Meteorol 188:64–75

Pavelsky TM, Smith LC (2004) Spatial and temporal patterns in Arctic river ice breakup observed with MODIS and AVHRR time series. Remote Sens Environ 93(3):328–338

Pesaresi M (2000) Texture analysis for urban pattern recognition using fine-resolution panchromatic satellite imagery. Geogr Environ Model 4(1):43–63

Prowse TD, Beltaos S (2002) Climatic control of river-ice hydrology: a review. Hydrol Process 16(4):805–822

Prowse TD, Conly FM (1998) Effects of climatic variability and flow regulation on ice-jam flooding of a northern delta. Hydrol Process 12(10–11):1589–1610

Rawlins MA, McDonald KC, Frolking S, Lammers RB, Fahnestock M, Kimball JS et al (2005) Remote sensing of snow thaw at the pan-Arctic scale using the SeaWinds scatterometer. J Hydrol 312(1–4):294–311

Sakai T, Hatta S, Okumura M, Hiyama T, Yamaguchi Y, Inoue G (2015) Use of Landsat TM/ETM+ to monitor the spatial and temporal extent of spring breakup floods in the Lena River, Siberia. Int J Remote Sens 36(3):719–733

Sakamoto T, Van Nguyen N, Kotera A, Ohno H, Ishitsuka N, Yokozawa M (2007) Detecting temporal changes in the extent of annual flooding within the Cambodia and the Vietnamese Mekong Delta from MODIS time-series imagery. Remote Sens Environ 109(3):295–313

Shulyakovskii LG (1972) On a model of the break-up process. Sov Hydrol Select Papers 1:21–27

Takakura H (2012) The local conceptualization of river ice thawing and the spring flood of Lena River under the global warming. Paper presented at the Proceedings of 1st International Conference Global Warming and the Human-Nature dimention in Siberia, pp 68–70

Vey S, Steffen H, Müller J, Boike J (2013) Inter-annual water mass variations from GRACE in central Siberia. J Geod 87(3):287–299

Wu J, Hobbs R (2002) Key issues and research priorities in landscape ecology: an idiosyncratic synthesis. Landsc Ecol 17(4):355–365

Xiao X, Boles S, Frolking S, Salas W, Moore B, Li C et al (2002) Observation of flooding and rice transplanting of paddy rice fields at the site to landscape scales in China using VEGETATION sensor data. Int J Remote Sens 23(15):3009–3022

Chapter 5
Detection of Ice-Jam Floods Using PALSAR Full-Polarimetry Data

Manabu Watanabe, Hiroki Takakura, Chinatsu Yonezawa, Yasuhiro Yoshikawa, and Masanobu Shimada

Abstract The Lena River in Russia at its intersection with the Arctic Ocean is subject to spring floods such as those caused by ice jams. In such cases, snowmelt occurs at lower latitudes and river water freezes at higher latitudes. That snowmelt is blocked by the frozen river water, which induces ice-jam floods every year. Full-polarimetric parameters obtained from the Phased Array Type L-band Synthetic Aperture Radar (PALSAR) were used to determine the most suitable ones for detecting ice-jam floods. Representative full-polarimetric parameters were calculated, including σ^0, entropy/α/anisotropy, the four-component decomposition model, and coherence. The PALSAR mosaic image helped to identify two large-scale floods, at Ytyk Kiuel on May 18, 2007, and at Berdigestiakh on July 15, 2007. These flood events were confirmed by articles published in the local newspaper and by optical sensor images. The optimum detectability for the flooding area was derived by the difference of polarimetric coherence, $\gamma_{(HH+VV)-(HH-VV)}$, obtained with and without the consideration of flooding. Total accuracy was 92.9%, and user and producer accuracies for detection of the flooded areas were 53.5% and 72.8%, respectively.

M. Watanabe (✉) • M. Shimada
Japan Aerospace Exploration Agency/Earth Observation Research Center, 2-1-1 Sengen, Tsukuba 305-8505, Japan

School of Science and Engineering, Tokyo Denki University, Ishizaka, Hatoyama, Hiki, Saitama 350-0394, Japan
e-mail: 16hz001@ms.dendai.ac.jp; shimada@g.dendai.ac.jp

H. Takakura
Center for Northeast Asian Studies, Tohoku University, 41 Kawauchi, Aoba-ku, Sendai 980-8576, Japan
e-mail: hrk@m.tohoku.ac.jp

C. Yonezawa
Graduate School of Agricultural Science, Tohoku University, 1-1 Amamiya-machi, Tsutsumidori, Aoba-ku, Sendai, Miyagi 981-8555, Japan
e-mail: chinatsu@m.tohoku.ac.jp

Y. Yoshikawa
Kitami Institute of Technology, Koenmachi 165, Kitami 090-8507, Japan
e-mail: yoshi@mail.kitami-it.ac.jp

T. Hiyama, H. Takakura (eds.), *Global Warming and Human - Nature Dimension in Northern Eurasia*, Global Environmental Studies,
DOI 10.1007/978-981-10-4648-3_5

The four-component decomposition analysis indicated that flood detection using $\gamma_{(HH + VV) - (HH - VV)}$ is applicable to low-height vegetation areas and urban regions.

Keywords L-band SAR • Disaster • Global warming

5.1 Introduction

The Lena River at its intersection with the Arctic Ocean is subject to spring floods. Snowmelt occurs at lower latitudes and the river freezes at higher latitudes. The time lag between these phenomena induces annual ice-jam flooding, which is a particular type of spring flood. Several reports have identified a change in scale and shift in the period of spring flooding. For the period 2005–2009, major spring floods on the Lena River were reported by the local newspaper in 2007 and 2008. One possible cause of frequent and large-scale flooding is global warming. Detecting the frequency and size of the spring floods in this region is important for examining the progression of global warming, and satellite observation is essential for this type of research.

Optical sensor data enables easy detection of flood magnitude in the absence of cloud cover. However, such cloud-free conditions occur in only 20% of satellite optical observations. Conversely, synthetic aperture radar (SAR) observation is available for all weather conditions day and night and is an effective tool for detecting flood size and frequency within a limited period.

Several studies have used SAR to detect flooding, the most popular and simple method of which is use of the radar backscattering coefficient (σ^0). SAR images include specular reflection of a water surface. If there is no radar backscatter, the water surface is represented as a darker image (Pierdicca et al. 2013). This is also valid for a vacant piece of land. An increase in σ^0 is observed for an inundated vegetation field because double-bounce scattering involving stems or trunks is enhanced by the specular reflection of floodwater (Pulvirenti et al. 2013).

Comparatively few studies have been conducted to detect flooding in urban areas because of the complexity of the radar-scattering mechanism. Even fewer studies involve a high-resolution global area of 30-m spatial resolution observed over 46 days. In this study, Phased Array Type L-band Synthetic Aperture Radar (PALSAR) full-polarimetry images were used to estimate suitable parameters for detecting floods, particularly for urban areas. Moreover, these images show the applicability of SAR data for monitoring spring floods.

Table 5.1 Observed data for mosaic image derived by Phased Array Type L-band Synthetic Aperture Radar (PALSAR)

Period	Off-nadir angle	Number of scenes
11: April 22–June 6, 2007	21.5°	194
27: April 27–June 11, 2009	23.1°	80

5.2 Data and Analysis

5.2.1 PALSAR Mosaic Image

PALSAR is an active microwave sensor onboard the Advanced Land Observing Satellite (ALOS) that uses L-band frequency. We used full-polarimetry data with a 30-m spatial resolution. ALOS had a unique basic observation plan in which global land areas were observed systematically during its operation (Rosenqvist et al. 2007). Our target area, the Lena River and surroundings, was observed in full-polarimetry mode during 2007 and 2009 in spring. Level 1.5 ground range amplitude data were collected to produce a mosaic image, as summarized in Table 5.1. To suppress total image size, spatial resolution was reduced from its original 30–50 m for that image. The polarimetric mosaic images (σ^0) obtained over 46 days in 2007 and 2009 are presented in Fig. 5.1. Red, green, and blue colors were assigned to σ^0_{HH}, σ^0_{HV}, and σ^0_{VV} polarizations, respectively. Latitudes and longitudes in the figure are 64.1°N, 119.6°E in the upper left corner to 60.0°N, 136.3°E in the lower right corner. There were few image gaps in the 2007 image with an off-nadir angle of 21.5°, because of its narrow swath (30 km) in full polarimetry mode. This lack of gaps was because the latitude of the target area is > 60°, where the satellite orbital interval is narrower than that at lower latitudes. However, there were several image gaps in the 2009 image, which had an off-nadir angle of 23.1° because the swath width was 23 km.

The green, bright purple, and dark purple colors in Fig. 5.1 represent forests, bare soil, and forests with frozen trees, respectively. σ^0 from frozen trees is 6–8 dB lower than that from non-frozen trees. The difference of σ^0 between frozen and non-frozen trees is larger with σ^0_{HV} polarization than with σ^0_{HH} (Watanabe et al. 2012), producing the dark purple color for the forest with frozen trees. Water areas are black in the figure because specular reflection causes only forward scattering; no radar backscatter was observed.

5.2.2 PALSAR Level 1.1 Complex Data Analysis

The mosaic image helped to identify two large-scale floods, at Ytyk Kiuel on May 18, 2007, and at Berdigestiakh on May15, 2007. These flood events were confirmed by articles published in the local newspaper (Fig. 5.2). Ytyk Kiuel was observed by

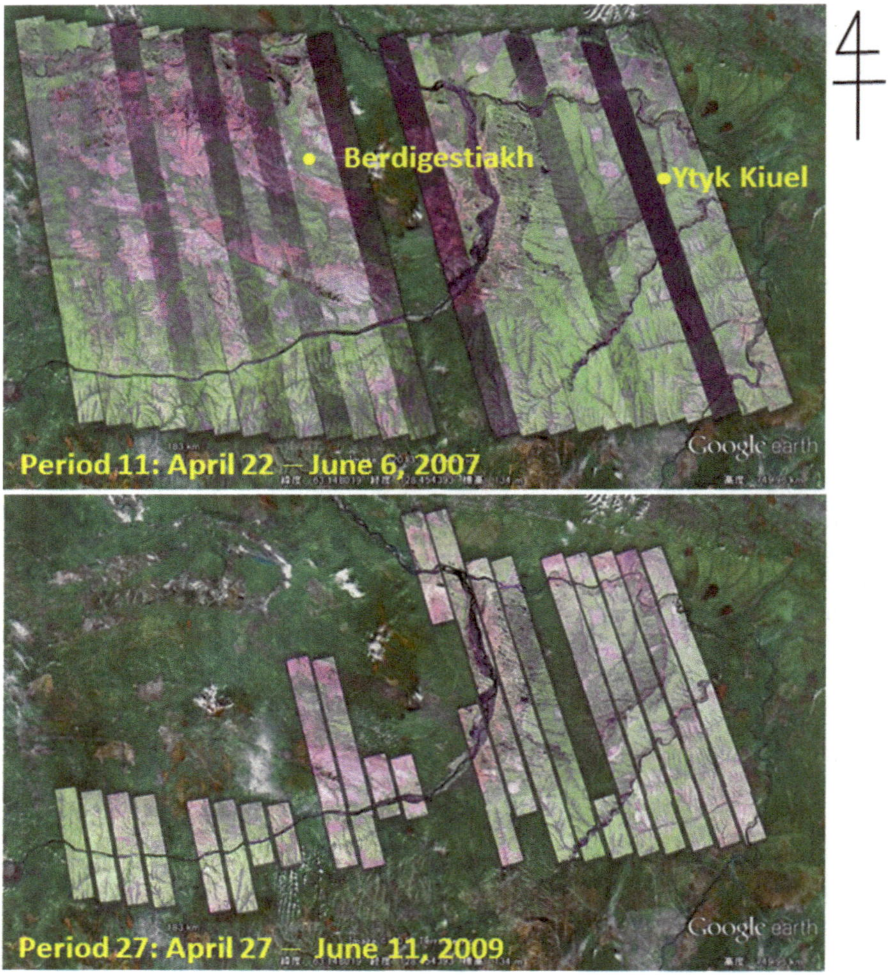

Fig. 5.1 Polarimetric mosaic image derived by PALSAR in periods 11 (April 22–June 6, 2007) and 27 (April 27–June 11, 2009). Red: HH polarization. Green: HV polarization. Blue: VV polarization

PALSAR on May 25, 2007, 1 week after the flooding, and on November 27, 2008 and May 1, 2009 when no flooding was recorded (Table 5.2). Berdigestiakh was observed by PALSAR on May 16, 2007, 1 day after the flooding. The same areas were observed on June 7, 2009, when no flooding was recorded. Level 1.1 calibrated slant range complex data (Shimada et al. 2009) were processed by PolSARpro (Pottier et al. 2009) and self-produced software. Several representative polarimetric parameters were calculated, including σ^0_{HH}, σ^0_{HV}, σ^0_{VV}, σ^0_{HH+VV}, σ^0_{HH-VV}, eigenvalue decomposition (Cloude and Pottier 1996), coherence (γ_{HH-VV}, $\gamma_{(HH+VV)-(HH-VV)}$,

Fig. 5.2 Newspaper article describing ice-jam flooding on May 18, 2007, in Ytyk Kiuel, Russia

Table 5.2 Satellite observations of ice-jam flooding

	Ytyk Kiuel	Berdigestiakh
April 25, 2007	*AVNIR-2*	
May 3, 2007		***AVNIR-2***
May 12, 2007	*AVNIR-2*	
May 15, 2007		**(Flooding)** AVNIR-2
May 16, 2007		PALSAR 21.5°
May 18, 2007	**(Flooding)**	
May 25, 2007	PALSAR 21.5°	
May 29, 2007	AVNIR-2	
Aug. 12, 2007	*PRISM*	
Nov. 27, 2008	*PALSAR 21.5°*	
April 20, 2009	***AVNIR-2***	
May 1, 2009	*PALSAR 23.1°*	
May 17, 2009	*AVNIR-2*	
June 7, 2009		*PALSAR 23.1°*
May 23, 2010		***AVNIR-2***

Italics text represents absence of flooding
Bold italics text indicates interference by cloud cover

γ_{RR-LL}), and the four-component decomposition model (Yamaguchi et al. 2005). The eigenvalue decomposition parameters entropy, α, and anisotropy were calculated from the eigenvalue and eigenvector obtained from the polarimetric data. Entropy represents the complexity of the scattering mechanism. α represents the scattering mechanism, where 0°, 45°, and 90° indicate surface scattering, scattering from a wire, and double-bounce scattering, respectively. Anisotropy represents the relative importance of the second and third eigenvalues. γ symbolizes the correlation coefficient between two polarizations. The four-component decomposition consists of four parameters related to surface, volume, double-bounce, and helix-scattering components on the earth surface.

The preferred parameters for flood detection were determined from among the calculated parameters. The derived slant range images were orthorectified using PolSARpro software. The flood timing images from 2007 were divided or subtracted by non-flood timing images of 2008–2009 to emphasize their differences. The orthorectified images were also converted into KML files so they could be superimposed on Google Earth images.

5.2.3 Optical Sensor Data

Data were collected using Advanced Visible and Near Infrared Radiometer type 2 (AVNIR-2) and Panchromatic Remote-sensing Instrument for Stereo Mapping (PRISM) onboard the ALOS satellite to derive additional information about the flooded areas. AVNIR-2 has four bands and a 10-m spatial resolution. PRISM is a panchromatic radiometer with 2.5-m spatial resolution.

Five observations were made in spring at Ytyk Kiuel with AVNIR-2 (Table 5.2), one of which was on May 29, 2007, 11 days after the flood and 4 days after the PALSAR observation. Another observation was on May 12, 2007, 6 days before the flooding. Nearly the entire area was covered by clouds during observation on April 20, 2009. Data from the other two observations on April 25, 2007, and May 17, 2009, produced clear images without flooding. A digital elevation model (DEM) based on PRISM data from the area on August 12, 2007, was used to generate an orthorectified PALSAR image.

Three spring observations were made at Berdigestiakh with AVNIR-2 (Table 5.2), one of which was on May 15, 2007, the same day as the flooding indicated by the newspaper. Nearly the entire area was covered by clouds during the other two observations, on May 3, 2007 and May 23, 2010. An Advanced Spaceborne Thermal Emission and Reflection Radiometer global DEM (GDEM), a joint product of Japan's Ministry of International Trade and Industry and the U.S. National Aeronautics and Space Administration, was used to produce an orthorectified PALSAR image. A pixel spacing of 10 m was used to compare the PALSAR and optical data.

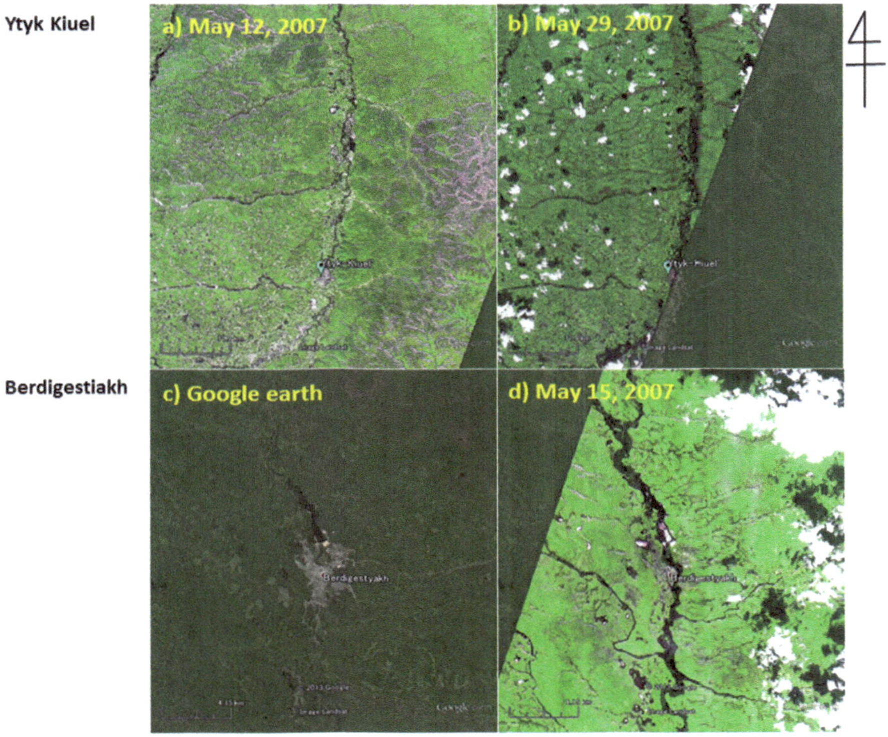

Fig. 5.3 AVNIR-2 natural color and Google Earth images derived under (**a**), (**c**) flooding and (**b**), (**d**) non-flooding conditions in Ytyk Kiuel and Berdigestiakh

5.3 Results and Discussion

5.3.1 Preferred Polarimetric Parameters for Flood Detection

AVNIR-2 natural color and Google Earth images derived for flooding and non-flooding conditions at Ytyk Kiuel and Berdigestiakh are presented in Fig. 5.3. The optical images with flooding conditions show a wider river than those with no flooding. Figure 5.4 shows the results of the four-component decomposition model in which red, green, and blue colors are assigned to double-bounce, volume, and surface scattering, respectively. These results also show a wider river under flooding conditions. In addition, there is a wide, bright red area along the river in flood shown by Fig. 5.4b, d). Optical sensor and Google Earth images (Fig. 5.3b, d) show that the area is covered by low-height vegetation. Thus, the bright red color may be attributed to stronger double-bounce scattering between the low-height vegetation and water underneath it. Strong backscatter was observed by the L-band SAR when the water was covered by vegetation (Shimada et al. 2010). Such conditions induced the

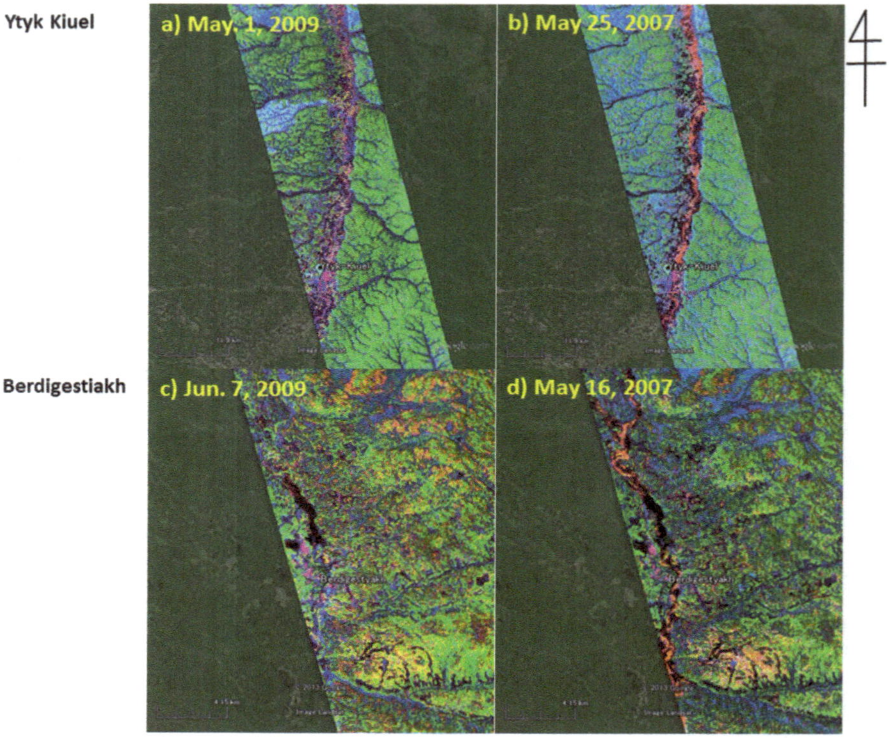

Fig. 5.4 Four-component decomposition model results under (**a**), (**c**) non-flooding conditions and (**b**), (**d**) flooding conditions. Red: double-bounce scattering. Green: volume scattering. Blue: surface scattering

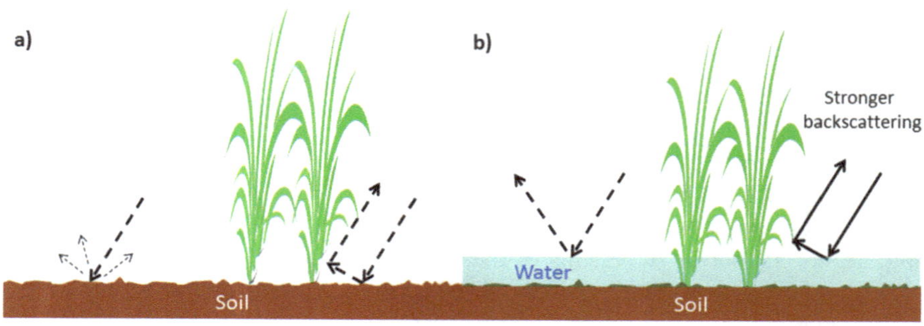

Fig. 5.5 Radar scattering from the ground under conditions (**a**) without water and (**b**) with water

stronger double-bounce scattering (Fig. 5.5) represented by the red color in Fig. 5.4. In urban areas, there was strong double-bounce scattering under both flooding and non-flooding conditions, and there was no clear difference between the two conditions. Two full-polarimetric parameters derived with PALSAR for 2007, entropy

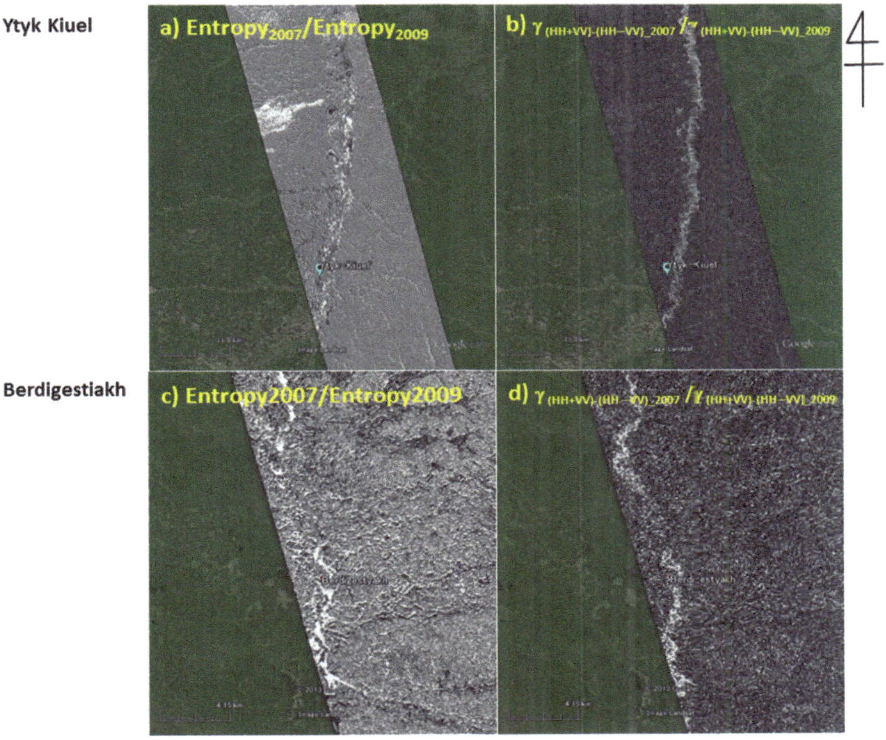

Fig. 5.6 Difference in entropy and $\gamma_{(HH + VV) - (HH - VV)}$ between flooding and non-flooding conditions

and $\gamma_{(HH + VV) - (HH - VV)}$, were divided by those obtained in 2009; the results are presented in Fig. 5.6. The entropy difference showed bright colors in areas where the surface conditions changed from water in 2007 to bare soil in 2009. The $\gamma_{(HH + VV) - (HH - VV)}$ parameter also highlights the differences between the flooding conditions in 2007 and non-flooding conditions in 2009.

Several other parameters such as the helix component of the four-component decomposition and double-bounce component of the Pauli decomposition showed similar results, although the contrast was weaker than that of entropy and $\gamma_{(HH + VV) - (HH - VV)}$.

5.3.2 Quantitative Evaluation

The detectability of flooded areas was evaluated using two of the prospective polarimetric parameters, entropy and $\gamma_{(HH + VV) - (HH - VV)}$. Both were calculated with several window sizes, i.e., 3, 7, 11, and 15 pixels with spacing 30 m/pixel Parameter values

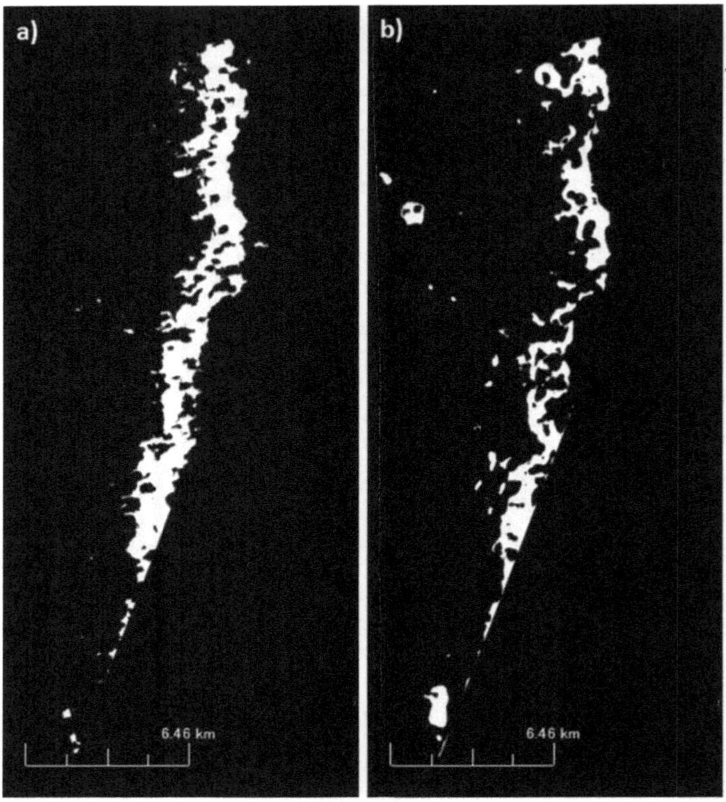

Fig. 5.7 (**a**) Pixels where $\gamma_{(HH+VV)-(HH-VV),2007} - \gamma_{(HH+VV)-(HH-VV),2009} \geq 0.2$ with window size of 15 pixels in Ytyk Kiuel. (**b**) Pixels where digital number of AVNIR-2 infrared band (band-4) is less than 24 on May 29, 2007 (after the flooding) and more than 25 on May 12, 2007 (before the flooding)

from 2009 were subtracted from those in 2007 to visualize the difference, with a parameter range of 0.1–0.5 and step of 0.1. The $\gamma_{(HH+VV)-(HH-VV)}$ results with window size 15 pixels at Ytyk Kiuel are presented in Fig. 5.7a. The detected areas totaled 27.1 km^2 and were distributed around the river. The same analysis was performed on the data obtained between 2008 and 2009, which showed no flooding. A small area of 0.1 km^2 was detected, indicating that no flooding occurred in 2008.

AVNIR-2 data were used for validation. Band 4 (infrared) was used to detect water-covered areas. To match the spatial resolution of PALSAR data, pixel averaging was done with window sizes 5, 9, 11, 17, 21, 27, and 31 pixels and 10 m/pixel spacing. Several threshold pixel numbers x of band 4, ranging from 5 to 25, were tested to detect the water areas. First, pixels at which the band 4 value was less than x on May 29, 2007 (after the flooding) were designated as flooded areas. Then, pixels at which the value was greater than $x + 1$ on May 12, 2007 (before the flooding)

Table 5.3 User and producer accuracies of flooding/non-flooding detection and window sizes for the analysis

	Ytyk Kiuel						Berdigestiakh		
	$\gamma_{(HH+VV)-(HH-VV),2007} - \gamma_{(HH+VV)-(HH-VV),2009}$[a]			Entropy2007 – $Entropy_{2009}$[b]			$\gamma_{(HH+VV)-(HH-VV),2007} - \gamma_{(HH+VV)-(HH-VV),2009}$		
	Flooding	Non-flooding	Producer's accuracy	Flooding	Non-flooding	Producer's accuracy	Flooding	Non-flooding	Producer's accuracy
Flooding	145402	54457	**72.8**	5243	208758	**2.4**	93390	69788	**57.2**
Non-flooding	126545	2232101	94.6	1418	2341447	99.9	39002	1795450	97.9
User's accuracy	**53.5**	97.6	92.9	**78.7**	91.8	91.8	**70.5**	96.3	94.6
Window size (PALSAR)	15			15			15		
Window size (AVNIR-2)	31			21			21		
Water area threshold	24			25			25		

Bold text represents user and producer accuracies for flooded area detection
[a]Difference $\gamma_{(HH+VV)-(HH-VV)}$, obtained under flooding and non-flooding conditions
[b]Difference of entropy, obtained under flooding and non-flooding conditions

Table 5.4 Four-component decomposition analysis results for areas detected as flooded in 2007, as shown in Fig. 5.7b

	Scattering component	Power (dB)
May 25, 2007	Double	−9.0
	Volume	−8.2
	Surface	−7.6
	Helix	−23.3
	Total	−3.4
May 1, 2009	Double	−8.9
	Volume	−3.4
	Surface	−2.1
	Helix	−20.0
	Total	−0.8

were assigned as non-flooding areas. By applying these two filters, the water area after flooding was determined. One of the results, with an averaging window size of 31 and threshold value of 24, is presented in Fig. 5.7b. Areas detected with patterns resembling $\gamma_{(HH+VV)-(HH-VV)}$ were also distributed around the river. The best results, which showed substantial accuracy in flooded area detection, are presented in Table 5.3. User and producer accuracies were evaluated from the error matrices to estimate classification accuracy (Story and Congalton 1986). User accuracy is a measure of the probability that a pixel is Class A, given that the classifier has labeled the pixel as that class. Producer accuracy is a measure of the probability that the classifier has labeled an image pixel as Class A, given that the ground truth is Class A. The best results, showing favorable accuracy in flooded area detection, are presented in Table 5.3.

The overall classification accuracy was > 90% for both $\gamma_{(HH+VV)-(HH-VV)}$ and entropy. However, respective detection accuracies of the flooded area showed user and producer accuracies of 53.5% and 72.8% for $\gamma_{(HH+VV)-(HH-VV)}$, and 78.7% and 2.4% for entropy. Thus, $\gamma_{(HH+VV)-(HH-VV)}$ produced better results.

No AVNIR-2 data were available for non-flooding conditions in the Berdigestiakh case. Therefore, accuracies were estimated from areas of both flooding and river. Results are listed in Table 5.3. Total accuracy was 94.6%, and user and producer accuracies of flood area detection were 70.5% and 57.2%, respectively. The difference $\gamma_{(HH+VV)-(HH-VV)}$ for flooding and non-flooding conditions also detected the flooded areas well.

5.3.3 Four-Component Decomposition Analysis

To obtain land cover information, four-component decomposition analysis was applied to the areas detected by AVNIR-2 as flooded (Fig. 5.7b). Results are summarized in Table 5.4. Double-bounce scattering for flooding conditions in 2007

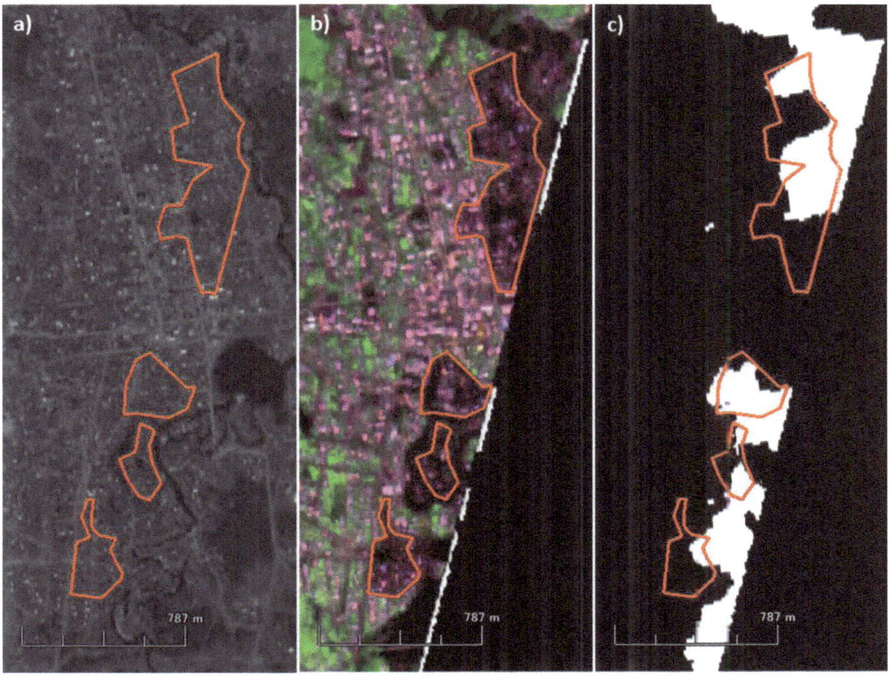

Fig. 5.8 (**a**) PRISM image for Ytyk Kiuel (August 12, 2007) (**b**) AVNIR-2 image for Ytyk Kiuel during flooding conditions (May 29, 2007) (**c**) Magnified images of Fig. 5.7(a)

appeared nearly identical to that under non-flooding conditions in 2009. However, other components under flooding conditions showed smaller values than those for non-flooding, which may be because the water surface suppress the surface and volume scattering. This result supports the fact that the difference $\gamma_{(HH+VV)-(HH-VV)}$ effectively detects areas of water-inundated vegetation of low height (Fig. 5.7b). $\gamma_{(HH+VV)-(HH-VV)}$ indicates the coherence between single- and double-bounce scattering represented by the (HH + VV) and (HH − VV) components, respectively. The larger $\gamma_{(HH+VV)-(HH-VV)}$ in the flooded condition may be attributed to stronger forward scattering on the water surface than that on the bare soil surface, and a smaller contribution from volume and surface scattering. The value of $\gamma_{(HH+VV)-(HH-VV)}$ also indicates differences in inundated urban areas (Fig. 5.8c). Because buildings also produce double-bounce scattering, the detection mechanism in urban areas is similar to that in areas of low-height vegetation.

5.4 Summary and Conclusions

PALSAR full-polarimetry data with and without flooding were used to examine optimal parameters for detecting flooded areas, particularly urban ones. We used 2007 observations of the Lena River in Russia during a period of large-scale flooding, and 2009 observations when no such flooding was reported. Several representative full-polarimetric parameters were calculated, such as entropy/α/anisotropy, the four-component decomposition model, and polarimetric coherence. A mosaic image observed over 46 days helped to identify two large-scale floods, at Ytyk Kiuel on May 18, 2007, and at Berdigestiakh on July 15, 2007. These flood events were confirmed by reports published in a local newspaper and by optical sensor images (AVNIR-2). The greatest detectability of flooded area was derived by the difference $\gamma_{(HH+VV)-(HH-VV)}$, obtained with and without flooding. Total accuracy was 92.9%, and user and producer accuracies of the detection of the flooded areas were 53.5% and 72.8%, respectively. Four-component decomposition analysis indicated that the flood detection with $\gamma_{(HH+VV)-(HH-VV)}$ is applicable to both low-vegetation and urban areas.

Thus, the preferred parameters to detect flooding with L-band SAR are as follows.

- σ^0 decrease for vacant piece of land
- $\gamma_{(HH+VV)-(HH-VV)}$ increase or σ^0 increase in low-vegetation areas
- $\gamma_{(HH+VV)-(HH-VV)}$ increase in urban areas

In summary, our comprehensive study identified the preferred parameters for detecting flooded urban areas, and demonstrates the applicability of SAR data in monitoring spring floods.

Acknowledgments This work was supported by the Japan Society for the Promotion of Science (JSPS) Grants-in-Aid for Scientific Research, (B) 22404001 and (B) 22310148.

References

Cloude SR, Pottier E (1996) A review of target decomposition theorems in radar polarimetry. IEEE Trans Geosci Remote Sens 34(2):498–518

Pierdicca N, Pulvirenti L, Chini M, Guerriero L, Candela L (2013) Observing floods from space: experience gained from COSMO–SkyMed observations. Acta Astronaut 84:122–133

Pulvirenti L, Pierdicca N, Chini M, Guerriero L (2013) Monitoring flood evolution in vegetated areas using COSMOSkyMed Data: the Tuscany 2009 case study. IEEE J Sel Topics Appl Earth Observ Remote Sens 6(4):1807–1816

Pottier E, Ferro-Famil L, Allain S, Cloude S, Hajnsek I, Papathanassiou K, Moreira A, Williams M, Minchella A, Lavalle M, Desnos Y (2009) Overview of the PolSARpro V4.0 software. The open-source toolbox for polarimetric and interferometric polarimetric SAR data processing. Proc IEEE IGARSS 4:936–939

Rosenqvist A, Shimada M, Ito N, Watanabe M (2007) ALOS PALSAR: a pathfinder mission for global-scale monitoring of the environment. IEEE Trans Geosci Remote Sens 45(11):3307–3316

Shimada M, Isoguchi O, Tadono T, Isono K (2009) PALSAR radiometric and geometric calibration. IEEE Trans GRS 47(12):3915–3932

Shimada M, Tadono T, Rosenqvist A (2010) Advanced land observing satellite (ALOS) and monitoring global environmental change. Proc IEEE 98(5):780–799

Story M, Congalton RG (1986) Accuracy assessment: a user's perspective. Photogramm Eng Remote Sens 52(3):397–399

Watanabe M, Takakura H, Yonezawa C, Yoshikawa Y, Shimada M (2012) Estimation of spring flooding factor in sub-arctic regions using PALSAR full polarimetry mode. In: Proceedings of the 53rd autumn conference of the Remote Sensing Society of Japan, pp 243–244 (in Japanese)

Yamaguchi Y, Yajima Y, Yamada H (2005) Four-component scattering model for polarimetric SAR image decomposition. IEEE Trans Geosci Remote Sens 43:1699–1706

Chapter 6
Using Air Temperature Data to Calculate Changes in Ice Sheet Thickness on the Lena River to Predict Ice-Jam Disasters

Yasuhiro Yoshikawa, Hiroki Takakura, Manabu Watanabe, Tetsuya Hiyama, and Toru Sakai

Abstract Ice jams are common winter hazards in rivers. In this study, ice-jam-related disasters were predicted at fixed points on a river based on air temperature data alone to develop a simple prediction method. Specifically, data were collected regarding disasters caused by ice jams in two major municipalities along the Lena River of Russia, Lensk and Yakutsk. Ice-sheet thickness was calculated by an equation that takes air temperature as the sole input variable over a period from the beginning of ice sheet formation to complete ice sheet melting. The method for predicting disasters focuses on maximum ice sheet thickness and the daily rate of decrease in ice sheet thickness. Comparison between measured and calculated values indicate that the equation for calculating the ice sheet thickness is applicable to the Lena River. Disasters from ice jams at Lensk and Yakutsk were retrospectively predicted by calculating ice sheet thickness with the equation in which only air temperature was substituted. The calculation results suggest that ice jams cause disasters when air temperature is not high during ice sheet melting. The daily decrease of ice sheet thickness is slow under low air temperatures, so the ice sheet can persist over prolonged periods in the river channel. It is also suggested that a large value of maximum ice sheet thickness affects the likelihood of disasters caused by ice jams.

Y. Yoshikawa (✉)
Kitami Institute of Technology, 165, Koen-cho, Kitami 090-8507, Japan
e-mail: yoshi@mail.kitami-it.ac.jp

H. Takakura
Center for Northeast Asian Studies, Tohoku University, Kawauchi 41, Aobaku, Sendai 980-8576, Japan

M. Watanabe
Japan Aerospace Exploration Agency, Tsukuba 305-8505, Japan

T. Hiyama
Institute for Space-Earth Environmental Research, Nagoya University, Nagoya 464-8601, Aichi, Japan

T. Sakai
National Institute for Environmental Studies, 2-16 Onogawa, Tsukuba 305-8506, Japan

T. Hiyama, H. Takakura (eds.), *Global Warming and Human - Nature Dimension in Northern Eurasia*, Global Environmental Studies,
DOI 10.1007/978-981-10-4648-3_6

Keywords Ice jams • Ice sheet thickness • Air temperature • Lena River • Lensk • Yakutsk

6.1 Introduction

In river channels in cold regions, ice forms when air temperature is low, particularly when flow velocity decreases. River ice consists of hard ice sheets on the water surface, soft frazil ice in flowing water, and snow accumulation on the ice sheets. When air temperature and river discharge increase, ice in the river begins to melt, break up and flow downstream. Ice floes flowing downstream can block the river channel where it narrows or meanders and where there are bridge piers. When river water backs up at such locations, ice jams form and cause a sudden rise in water level. In cold regions, river floods are caused by rainfall, snowmelt, or ice jams. Ice jams occur in Japan (Yoshikawa et al. 2012) but the scale is much larger in other countries, such as the northern areas of the U.S., Canada, China and Russia, where winter temperatures also fall into the sub-zero range. A rise in water level from large-scale ice jams in these countries can directly lead to a disaster in which river water and ice inundate residential areas.

River freezing is closely linked to the lives of people living in the Lena River basin of Russia. In lakes, which become part of floodplains and are connected to the river channel during a flood, these people collect ice for drinking water and set baskets under ice sheets to catch small fish (Takakura 2013). They also use the frozen surfaces of rivers, lakes and wetlands as traffic routes (Okumura et al. 2011). However, large-scale ice jams during the ice-melt period cause great harm to people. In May 2001, there were large-scale ice jams in Lensk, with a population of 28,000 (Shahramanjan 2004). The water level increased to 12 m and the town was inundated. The town suffered four billion rubles in losses. Damage in the Sakha Republic as a whole amounted to six billion rubles.

One study has shown that dropping explosives from helicopters can be effective at breaking up ice jams (Shahramanjan 2004). Hydraulic experiments on ice jams in the Lena River (Buzin et al. 2007) suggest three measures for protecting Lensk from ice-jam-related disasters: (1) Controlling the discharge at a site 40 km downstream from Lensk, (2) breaking river ice to reduce its strength; (3) artificially forming ice jams at a site 20 km upstream from Lensk. According to the researchers who conducted the experiments, the creation of artificial ice jams is the most effective disaster countermeasure (Buzin et al. 2007). Other potentially effective countermeasures toward achieving the discharge of a greater quantity of ice and artificial breakup of ice jams include river improvement work, such as riverbed scouring and river channel widening.

In a study on the prediction of ice jams (Buzin and Kopaliani 2008), ice jam disasters in Lensk were examined using the water level 2 days prior to such events at a site 185 km upstream from that city. That study indicated the necessity for increasing prediction accuracy by developing a technique that takes into account factors such as heat balance, river ice strength, and river hydraulic characteristics.

It was also noted that the use of satellite data is effective for short-term forecasting of ice jams.

Simple techniques for accurate prediction as to whether ice jams will cause a disaster are required for river management in cold regions. Ice jams are complex phenomena that involve the formation and melting of river ice, its breakup by water flow, collisions with other river ice, and ice accumulation downstream. These phenomena are analyzed from the perspective of thermal and structural dynamics and hydraulics.

One study indicated that ice melt preceding ice jams was induced by increased air temperatures (Yoshikawa et al. 2012). In contrast, it was shown in a study focused on the Lena and other rivers (Pavelsky and Smith 2004) that there was little time lag between the onset of ice sheet melt and increase of water level because of ice jams. Thus, if we can predict when ice melts on the basis of air temperature data, we will also be able to assess the probability of disasters caused by ice jams.

In this chapter, we predict ice-jam-related disasters at fixed points on the river based on air temperature data alone to develop a simple prediction method. The validity of the equation used for this method was determined by research of the Teshio River in Hokkaido, Japan (Yoshikawa et al. 2014). However, the scales of the Teshio and Lena rivers are different. The validity of the equation for the Lena river was investigated by comparing observed and calculated values. Specifically, data regarding disasters caused by ice jams were collected in two major municipalities along the Lena River, namely, Lensk and Yakutsk (Fig. 6.1). Then, ice sheet thickness was calculated by the equation (Yoshikawa et al. 2014) that takes air tempera-

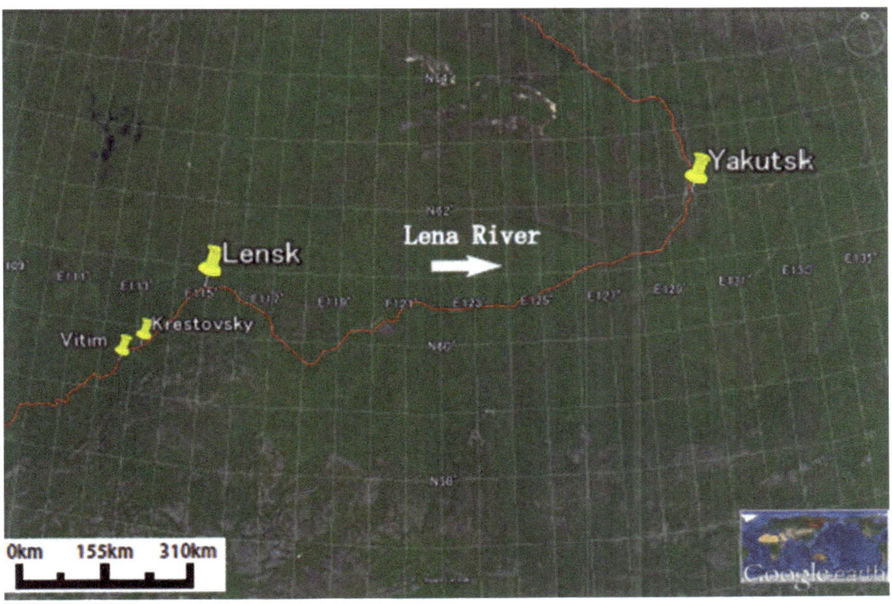

Fig. 6.1 Location of Lena River (Based on satellite photo obtained through Google Earth software)

ture as the sole input variable. Changes in ice sheet thickness were calculated over the period between the onset of ice sheet formation and final melting. A method for predicting disasters that focuses on the maximum ice sheet thickness and daily rate of decrease in ice sheet thickness was studied.

6.2 Applicability of Equation for Calculating Ice Sheet Thickness on Lena River

6.2.1 Equation for Calculating Ice Sheet Thickness (Yoshikawa et al. 2014)

Figure 6.2 shows a conceptual diagram of heat balance in a frozen river. In Fig. 6.2, $T[^\circ\mathrm{C}]$ is temperature, $h[\mathrm{m}]$ is a layer thickness, $\rho\left[\frac{\mathrm{kg}}{\mathrm{m}^3}\right]$ is density of the flowing water, frazil ice, ice sheets and snow cover, $U\left[\frac{\mathrm{m}}{\mathrm{s}}\right]$ is flow velocity, and $\phi\left[\frac{\mathrm{W}}{\mathrm{m}^2}\right]$ is heat flux. Subscripts indicate values of each layer.

Changes in snow depth are expressed by Eq. (6.1), which considers the heat balance between the air and ice sheets. In calculating changes of snow depth, this equation considers only changes owing to the heat balance, and not changes owing to snowfall or wind, which causes snow to drift or blow upward. Changes in ice sheet thickness are expressed by Eq. (6.2), which takes into account the heat balance

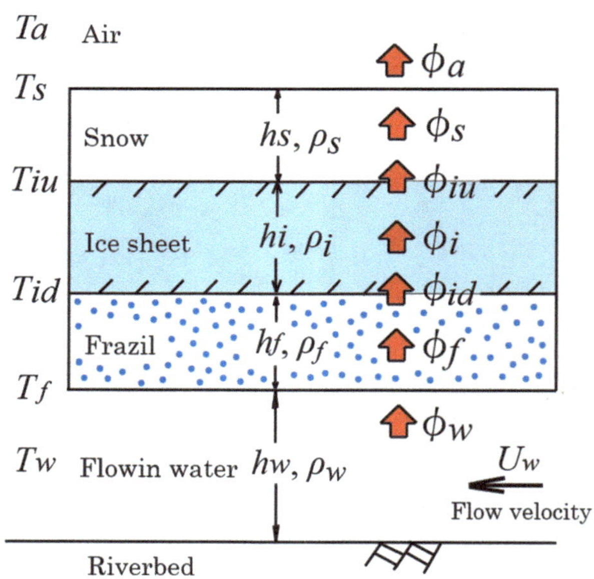

Fig. 6.2 Conceptual diagram of heat balance in a frozen river

between the snow cover and frazil ice. It is assumed that the interface between the snow cover and ice sheets, the interface between the ice sheets and frazil ice, and the interior of the ice sheets have variable heat flux values. Therefore, three different values of heat flux are used in Eq. (6.2). The heat flux is given by ϕ_{iu}, ϕ_{id}, and ϕ_i for the interface between the snow cover and ice sheets, the interface between the ice sheets and frazil ice, and the interior of the ice sheets, respectively. Changes in frazil ice thickness are expressed by Eq. (6.3), which considers the heat balance between ice sheets and flowing water. This equation accounts for changes in frazil ice thickness only due to heat balance. It does not consider increase of this thickness associated with frazil ice flowing from upper reaches of the river, nor does it consider decrease in the thickness caused by frazil ice flowing downstream.

Equations (6.1), (6.2) and (6.3) are rearranged into Eq. (6.4) as shown below.

$$\rho_s L_s \frac{dh_s}{dt} = (\phi_a - \phi_s) + (\phi_s - \phi_{iu}), \tag{6.1}$$

$$\rho_i L_i \frac{dh_i}{dt} = (\phi_{iu} - \phi_i) + (\phi_i - \phi_{id}), \tag{6.2}$$

$$\rho_f L_f \frac{dh_f}{dt} = (\phi_{id} - \phi_f) + (\phi_f - \phi_w), \tag{6.3}$$

$$\rho_i L_i \frac{dh_i}{dt} = \phi_a - \rho_s L_s \frac{dh_s}{dt} - \rho_f L_f \frac{dh_f}{dt} - \phi_w. \tag{6.4}$$

The heat flux in each layer is approximated by Eqs. (6.5) through (6.8) using the variables of heat exchange coefficient, thermal conductivity, ice layer thickness, and temperature difference.

$$\phi_a = h_{sa}(T_s - T_a), \tag{6.5}$$

$$\phi_s = \frac{k_s}{h_s}(T_{iu} - T_s), \tag{6.6}$$

$$\phi_i = \frac{k_i}{h_i}(T_{id} - T_{iu}), \tag{6.7}$$

$$\phi_f = \frac{k_f}{h_f}(T_f - T_{id}). \tag{6.8}$$

Assuming that the change at the interface of each layer at a given time is in equilibrium, then $\phi_a = \phi_s = \phi_i = \phi_f$. Thus, the heat flux from the river ice surface to the air ϕ_a is expressed by

$$\phi_a = \frac{T_f - T_a}{\dfrac{1}{h_{sa}} + \dfrac{h_s}{k_s} + \dfrac{h_i}{k_i} + \dfrac{h_f}{k_f}}. \quad (6.9)$$

The heat flux from flowing water to the bottom surface of river ice ϕ_w is expressed by Eq. (6.10) (Ashton 1986). C_{wi} is $1622 \frac{\mathrm{W \cdot s^{0.8}}}{^{\circ}\mathrm{C \cdot m^{2.6}}}$, $U_w \left[\frac{\mathrm{m}}{\mathrm{s}}\right]$ is the mean vertical flow velocity, and $h_w[\mathrm{m}]$ is the effective water depth from the riverbed to underside of the ice sheet.

$$\phi_w = C_{wi} \frac{U_w^{4/5}}{h_w^{1/5}} \times (T_w - T_f). \quad (6.10)$$

Manning's equation is typically used for calculating mean flow velocity in an iron or concrete pipe conduit with large cross-sectional flow area. The conditions of a frozen river channel are different from those of a pipe conduit because the pressure of river water is relieved by cracks in the river ice. Even so, the river surface is almost entirely covered with ice, so it is possible to assume that the frozen river channel is similar to a pipe conduit and that Manning's equation is applicable. That equation is used to obtain Eq. (6.11). Equation (6.12) is obtained by substituting Eq. (6.11) into Eq. (6.10). Here, n_c is Manning's roughness coefficient that deals with the combined roughness of the riverbed and river ice, and i is the hydraulic gradient.

$$U_w = \beta h_w^{2/3}, \beta = \frac{1}{2^{2/3}} \frac{\sqrt{i}}{n_c}, \quad (6.11)$$

$$\phi_w = C_{wi} \beta^{4/5} T_w h_w^{1/3}. \quad (6.12)$$

When a difference method is applied to Eq. (6.4) and when Eqs. (6.9) and (6.12) are substituted into it, Eq. (6.13) for ice sheet thickness is derived. Coefficient α is defined by Eq. (6.14), and 0 °C is given as the temperature at the underside of the river ice T_f.

$$h_i = h_i' - A \frac{T_a}{h_i'} - W T_w h_w^{1/3}, \quad (6.13)$$

$$\begin{cases} A = \left(\dfrac{k_i \Delta t}{\rho_i L_i}\right) \alpha \\ W = \left(\dfrac{C_{wi} \Delta t}{\rho_i L_i}\right) \beta^{4/5} \end{cases}$$

Table 6.1 Physical properties of ice

	h_{sa}	ρ_i	L_i	k_i	Δt
	Heat exchange coefficient	Density	Latent heat	Thermal conductivity	Time
Unit	$\frac{W}{m^2 \cdot {}^\circ C}$	$\frac{kg}{m^3}$	$\frac{W \cdot s}{kg} = \frac{J}{kg}$	$\frac{W}{m \cdot {}^\circ C}$	s
Value	25.0	917.4	3.336×10^5	2.31	$24 \times 60 \times 60$

$$
\begin{cases}
\alpha = \alpha' \times \alpha'' \\
\alpha' = 1 - \dfrac{\rho_s L_s \dfrac{dh_s}{dt}}{\phi_a} - \dfrac{\rho_f L_f \dfrac{dh_f}{dt}}{\phi_a} \\
\alpha'' = \dfrac{\dfrac{h_i'}{k_i}}{\dfrac{1}{h_{sa}} + \dfrac{h_s'}{k_s} + \dfrac{h_i'}{k_i} + \dfrac{h_f'}{k_f}}
\end{cases}
\tag{6.14}
$$

By substituting the values shown in Table 6.1 into Eq. (6.13), Eq. (6.15) is derived for calculating ice sheet thickness in a way that is more practical and simple than before.

$$
h_i = h_i' - \left(\frac{65.2}{10^5}\right)\alpha \frac{T_a}{h_i'} - \left(\frac{45.8}{10^2}\right)\beta^{4/5} T_w h_w^{1/3}. \tag{6.15}
$$

Equation (6.15) is explained as follows, using an example in which ice sheet thickness is calculated hourly. Because T_a and T_w in that equation are daily mean values, when T_a and T_w are needed as hourly values (i.e., 60×60 sec), they are respectively expressed as T_{a1} and T_{w1}, and these input values are expressed by $T_{a1} \times \frac{60 \times 60}{24 \times 60 \times 60}$ and $T_{w1} \times \frac{60 \times 60}{24 \times 60 \times 60}$. The ice sheet thickness h_i'[m] is given an initial value of 1 mm. It is subsequently assigned the value of 1 h, prior to calculation.

6.2.2 Calculation Conditions

The coefficient α (dimensionless) in Eq. (6.15) was determined by a trial-and-error method on the assumption that calculated values of ice sheet thickness accurately reproduce measured values. The value of α was determined to be 0.15. The value of $\beta\left[\frac{\mathrm{m}^{\frac{1}{3}}}{\mathrm{s}}\right]$ was obtained by Eq. (6.16).

$$\beta = \frac{U_w}{h_w^{2/3}}, U_w = \frac{Q}{Bh_w}, \tag{6.16}$$

$$h_w = \left(\frac{2^{2/3} n_c Q}{B\sqrt{I_b}}\right)^{3/5}. \tag{6.17}$$

At Krestovsky, 150 km upstream of Lensk (Fig. 6.1), discharge $Q\left[\mathrm{m}^3/\mathrm{s}\right]$ was determined to be 899, the mean discharge (ArcticRIMS) in the 6 months from November through April. Values of river width B (in m) and riverbed slope I_b (dimensionless) were determined as 815 m and 1/14833, respectively, based on data obtained using Google Earth software. The roughness coefficient $n_c\left[\frac{s}{\mathrm{m}^{\frac{1}{3}}}\right]$ was assigned a value of 0.03, typical for this coefficient.

Equation (6.18) was used for calculating water temperature (Yoshikawa et al. 2014).

$$T_w = \frac{T_a}{1+\gamma} + \frac{\gamma T_w'}{1+\gamma}, \tag{6.18}$$

$$\gamma = \frac{h_w \rho_w C_p}{(1-N)h_{wa}\Delta t}, \tag{6.19}$$

where $T_w[^\circ\mathrm{C}]$ is water temperature, $T_w'[^\circ\mathrm{C}]$ is water temperature at a time Δt before T_w, $\rho_w\left[\frac{kg}{m^3}\right]$ is water density (given a value of 999.84), $C_p\left[\frac{J}{kg\cdot{}^\circ\mathrm{C}}\right]$ is specific heat of water (given a value of 4200), $h_{wa}\left[\frac{W}{m^2\cdot{}^\circ\mathrm{C}}\right]$ is the heat exchange coefficient of the water surface (given a value of 20), $\Delta t[s]$ represents time intervals of calculation, and $N[dimensionless]$ is the cross-sectional river ice ratio. Specifically, N is the ratio of river ice width to river width B ($0 \leq N \leq 1$). N was determined to be 0.99 on the assumption that the river is entirely frozen when the mean transverse ice sheet thickness h_i is 0.7 m ($=h_{imax}$). Assuming that $N=0$ when $h_i=0$, N was determined to be directly proportional to the ice sheet thickness. Water temperatures $T_w < 0$ °C were regarded as 0 °C in the calculations. The daily mean temperature (Ohata et al. 2003) at Vitim, 200 km upstream of Lensk, was used for air temperature.

6.2.3 *Comparison of Measured and Calculated Values of Ice Sheet Thickness*

For verifying that calculated values of ice sheet thickness were consistent with measured ones, the former were compared with values measured every 10 days from November 1986 through April 1987 and November 1987 through April 1988 (Vuglinsky and Kubota 2003). Figures 6.3 and 6.4 show calculated and measured

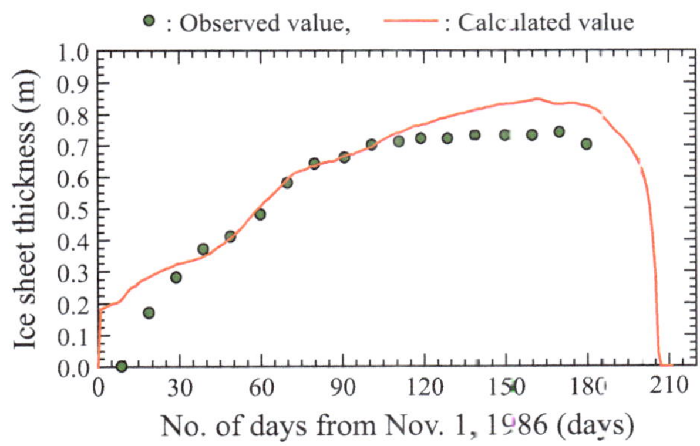

Fig. 6.3 Measured and calculated values of ice sheet thickness (1986–1987)

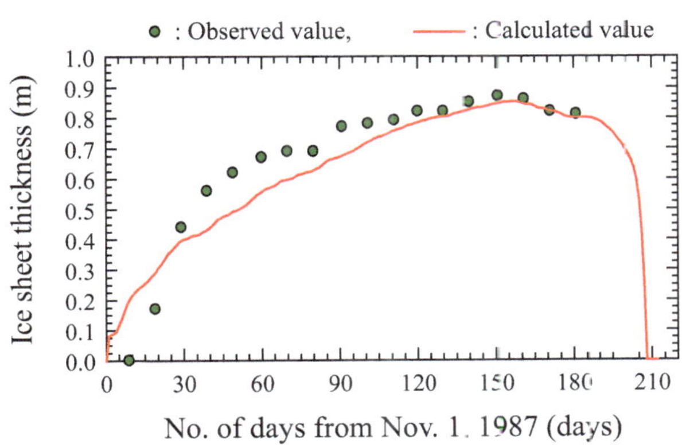

Fig. 6.4 Measured and calculated values of ice sheet thickness (1987–1988)

values for the 1986–1987 and 1987–1988 winters, respectively. Mean absolute error is 6.3 cm in Figs. 6.3 and 6.6 cm in Fig. 6.4. Both winter periods were accurately predicted by the calculated values. Thus, we conclude that the equation using air temperature alone as an input variable for predicting ice sheet thickness is valid for reproducing measured values.

6.3 Disasters Caused by Ice Jams Predicted Retrospectively Using Equation for Ice Sheet Thickness

Ice jams that caused damage at Lensk and Yakutsk on the Lena River were identified in a database created by Dartmouth College. From 1998 through 2009, ice jams were found to have caused damage in 1998, 1999, 2001, 2003 and 2007.

Figure 6.5 shows variations in calculated values of ice sheet thickness for each year during 1998 through 2009. These values were calculated using the daily mean temperature obtained at Vitim. Solid lines indicate years when ice jams caused damage. The x-axis shows days elapsed from September 1 of the previous year. For 1998, for example, the days are counted from September 1, 1997. Figure 6.5 clearly shows that the ice sheet thickness was calculated for the period between formation and melting. In addition, the freezing indexes in 1999 and 2001, with thick ice sheets, were −4109 °C and −4486 °C, respectively. The indexes in 2002 and 2008, with thin ice sheets, were −2816 °C and −2753 °C. The average index for 1998–2009 was –3540 °C. Air temperature and ice sheet thickness are positively correlated.

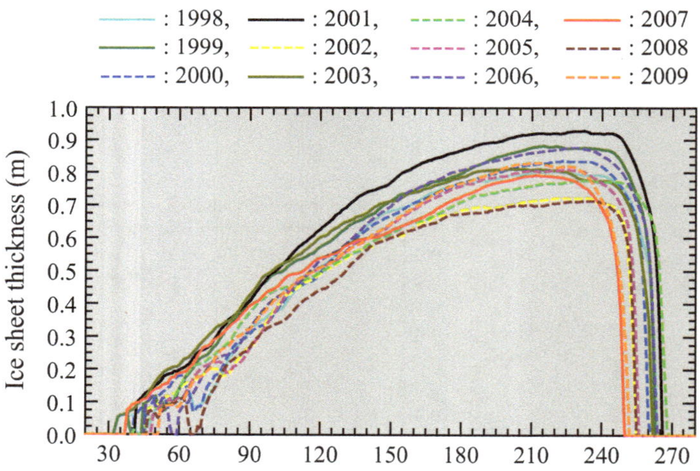

Fig. 6.5 Calculated values of ice sheet thickness on Lena River at Vitim (years with disasters caused by ice jams shown by *solid lines*)

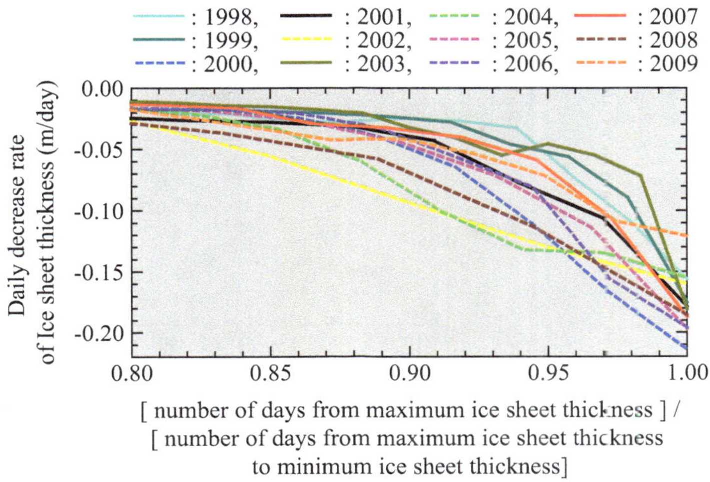

Fig. 6.6 [Number of days from maximum ice sheet thickness]/[number of days from maximum ice sheet thickness to minimum ice sheet thickness], shown by x-axis; y-axis shows daily rate of decrease in ice sheet thickness

In Fig. 6.6, the x-axis shows the [number of days from maximum ice sheet thickness]/[number of days from maximum ice sheet thickness to minimum ice sheet thickness]. The value is 0 when the ice sheet thickness is maximum. The value is 1 when the ice sheet is completely melted. The y-axis shows the daily rate of decrease in ice sheet thickness. The figure shows that the daily rate of decrease was slower in years with ice-jam induced disasters. In the 0.8–1.0 range of the x-axis in the figure, the average rate of daily decrease for years with ice jams was −1.42 cm/day. That rate for years without ice jams was −2.16 cm/day. Years with ice jams had smaller daily rates of decrease in ice sheet thickness. In the ice-melt period, an increase in air temperature melts river ice and increases river flow speed. Thus, increased air temperature is a major influence on ice melt. The calculation results suggest that ice jams cause disasters when air temperature does not become high and ice persists in the river channel for a prolonged period under low temperature.

Ice jams produced a large-scale disaster in 2001, during which the daily rate of decrease in ice sheet thickness (y-axis, black solid line in Fig. 6.6) was faster than those in other years with ice-jam disasters. That rate in 2001 is similar to those in years without such disasters. In Fig. 6.7, the y-axis indicates maximum ice sheet thickness and the x-axis the number of days from maximum ice sheet thickness to minimum ice sheet thickness. Maximum ice sheet thickness was greater in 2001 than in any other year (Figs. 6.5 and 6.7). In addition to a slower daily rate of decrease in ice sheet thickness during the ice-melt period, a large maximum thickness is associated with a greater likelihood of ice-jam-related disaster.

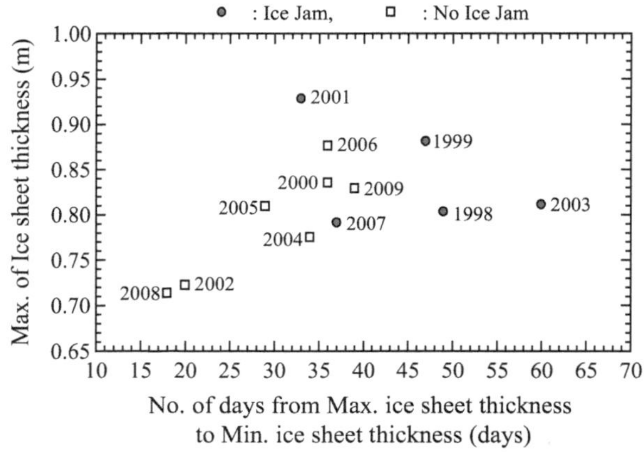

Fig. 6.7 Number of days from maximum ice sheet thickness to minimum ice sheet thickness, shown by x-axis; y-axis shows maximum ice sheet thickness

Above 0.98 on the x-axis in Fig. 6.6, the daily rate of decrease in ice sheet thickness is slower for 2009 than for 2007. A comparison of satellite synthetic aperture radar data (Watanabe et al. 2012) between 2007 and 2009 indicates that ice melt and air temperature increase in areas near forests occurred more slowly in 2009 than in 2007. This is consistent with the results of the present study. On the x-axis of Fig. 6.7, the number of days from maximum ice sheet thickness to minimum ice sheet thickness is larger for 2009 than 2007. There was no disaster caused by ice jams in 2009, even though conditions for such a disaster were satisfied as in years with disasters. Three assumptions can be considered for this finding: (1) The ice broke up on a small scale, so no ice jams formed; (2) ice jams developed in other locations; (3) countermeasures against ice jam formation were effective. We could not explain why 2009 was different from the other years regarding disaster occurrence.

6.4 Summary

Comparison between the measured and calculated values indicated that the equation used for ice sheet thickness can be used to make predictions for the Lena River. Disasters from ice jams at Lensk and Yakutsk along the Lena were retrospectively predicted by calculating ice sheet thickness with the equation, into which only air temperature was substituted. The calculation results suggest that ice jams cause disasters when air temperature does not become high during the ice-melt period. In such a case, the ice sheet thickness does not decrease and ice in the river channel can persist for a prolonged period. A large maximum ice sheet thickness appears to be another factor that affects the likelihood of disasters from ice jams.

Assessment of disaster produced by ice jams appears possible using the present method for predicting them. The method uses observed air temperature along the river for calculating ice sheet thickness.

Acknowledgments Shigemi Hatta of Tomakomai National College of Technology provided us with research data. The study was supported by a grant from the Foundation of River & Watershed Environment Management (River Improvement Fund 25-1212-002) and JSPS (Japan Society for the Promotion of Science) Kakenhi Grant-in-Aid for Young Scientists (B) 25870023. Their assistance and support are greatly appreciated.

References

ArcticRIMS A regional, intergrated hydrological monitoring system for the pan-Arctic land mass, December 2013. http://rims.unh.edu/
Ashton GD (ed) (1986) River lake ice engineering. Water Resources Publications, Colorado, p 289
Buzin VA, Klaven AB, Kopaliani ZD (2007) Laboratory modelling of ice jam floods on the Lena River, extreme hydrological events: new concepts for security. NATO science series, vol 78, pp 269–277
Buzin, VA, Kopaliani ZD (2008) Ice jam floods on the rivers of Russia: risks of their occurrence and forecasting, Ecwatech in Moscow, pp 582–587
Google Earth December 2013, http://earth.google.co.jp/
Ohata T, Razuvaev VN, Suzuki R (2003) Baseline meteorological data in Siberia, in dataset for water and energy cycle in Siberia (Version 1),GAME-Siberia and Frontier observational research system for global change
Okumura M, Kawamoto K, Boyakova S (2011) Winter roads and effects of global warming in the Sakha Republic of the Russian Federation. Trans Policy Stud 14(3):16–23
Pavelsky TM, Smith LC (2004) Spatial and temporal patterns in Arctic river ice breakup observed with MODIS and AVHRR time series. Remote Sens Environ 93:328–338
Shahramanjan MA (2004) Ice jams forecast and technologies of their destruction on the rivers of the Russian Federation. International Association of Hydraulic Engineering and Research, International Symposium on Ice, 17th, pp XXVII–XLII
Takakura H (2013) Is an Ice-jam Flood a Disaster? Indigenous knowledge of river ice and adaptation of the sakha communities in the Middle Basin of the Lana River. Northeast Asian Stud 17:109–138
Vuglinsky VS, Kubota J (2003) Hydrological and ice regime data of river in Eastern Siberia. In: Dataset for water and energy cycle in Siberia (Version 1), GAME-Siberia and Frontier Observational Research System for Global Change
Watanabe MH, Takakura C, Yonezawa Y, Yoshikawa Y, Shimada M (2012) Use of PALSAR full-polarimetric data for estimating the cause of floods in the Polar Region, The Remote Sensing Society of Japan, collection of papers presented at the 53rd conference of the RSSJ, pp 243–244
Yoshikawa Y, Watanabe Y, Hayakawa H, Hirai Y (2012) Field observation of a river ice jam in the Shokotsu River in February 2010. In: Proceedings of the 21st IAHR international symposium on ice, pp 105–117
Yoshikawa Y, Watanabe Y, ITOH A (2014) A simple equation for ice sheet thickness and ice formation/breakup prediction. J JSCE 2:203–213

Chapter 7
Ice Movement in the Lena River and Effects of Spring Flooding on Human Society: An Interpretation of Local Sources Integrated with Satellite Imagery in a Multidisciplinary Approach

Hiroki Takakura, Yasuhiro Yoshikawa, Manabu Watanabe, Toru Sakai, and Tetsuya Hiyama

Abstract Spring flooding of Arctic rivers is an important yet largely unexplored issue in climate change research. Owing to their complexity, scientific data on how climate change is altering the freshwater ice process and its impact on human societies remain unclear. This chapter explores the process of spring flooding on the Lena River in Siberia. We describe ice movement and speed, loci of flooding outbreaks, and the disastrous results. The analysis is an interpretative approach using multidisciplinary data from satellite imagery, local newspaper articles, and anthropological fieldwork. Combining the results of various disciplines, we obtained the following findings. Spring flooding occurs in conditions featuring a relatively slow speed of drifting ice. This flooding always occurs in the Lena River basin but, in years with dramatic water rise, there is damage even to surrounding regions. This more severe

H. Takakura (✉)
Center for Northeast Asian Studies, Tohoku University,
Kawauchi 41, Aobaku, Sendai 980-8576, Japan
e-mail: hrk@m.tohoku.ac.jp

Y. Yoshikawa
Kitami Institute of Technology, Kitami 090-8507, Japan
e-mail: yoshi@mail.kitami-it.ac.jp

M. Watanabe
Japan Aerospace Exploration Agency, Tsukuba 305-8505, Japan
e-mail: watanabe.manabu@jaxa.jp

T. Sakai
National Institute for Environmental Studies, 2-16 Onogawa, Tsukuba 305-8506, Japan
e-mail: sakai.toru@nies.go.jp

T. Hiyama
Institute for Space-Earth Environmental Research, Nagoya University,
Nagoya 464-8601, Aichi, Japan
e-mail: hiyama@nagoya-u.jp

T. Hiyama, H. Takakura (eds.), *Global Warming and Human - Nature Dimension in Northern Eurasia*, Global Environmental Studies,
DOI 10.1007/978-981-10-4648-3_7

flooding clearly causes damage that affects the population. However, in years with moderate ice or shorter ice-jam flooding, the effects are beneficial to local agriculture. The spring flooding is a double-edged sword and the effects on the human society should be evaluated from the perspective of the aforesaid complex interactions.

Keywords Ice-jam flood • Ice movement • Lena river • Local society • Human–river relationships

7.1 Introduction

Spring flooding of northern rivers in both Arctic and sub-arctic regions is a regular seasonal event. As snow and ice thaw, floating freshwater ice appears in the rivers and flooding occurs, which is critical for both the regional ecosystem and water circulation.

The spring flooding is sometimes an ice-jam flood, in which case it is the result of a high water level rather than an open-water flood caused by heavy rainfall or similar weather-related circumstances (Beltaos 1995: 17). The peculiar phenomena that compose the mechanism of ice-jam flooding are complex, so predicting these events and adapting to them is difficult (Beltaos 1995: 7–17; Yoshikawa et al. 2012). Moreover, these floods have a great impact on local human societies in terms of risk and cost (Prowse 2007: 207).

Many scientific data point to a change in scale and shift in the period of spring flooding. Long-term records (150 years) of river ice of the Northern Hemisphere show that a 2–3 °C increase in air temperature has produced an approximate 10- to 15-day delay in river freezing, and that the breakup of this ice has moved to earlier an earlier time in spring by about the same number of days (Prowse 2007: 204). Another source from 2005 (Huntington and Weller 2005) reported that compared with the 1950s, there was an average annual temperature increase of 2–3 °C, with one winter showing a 4 °C increase. The same study predicted an additional warming of 1 °C by 2020, 2–3 °C by 2050, and 4–5 °C by 2008.

Theoretically, climate change should have some effect on spring flooding of the Lena River, but owing to regional variations and the complex interplay of elements, scientific data on how the climate is altering the freshwater ice process (ice-cover composition, thickness, and dynamics of ice break-up) remain unclear (Prowse 2007: 202).

As an important and integral element of the river flow regime, river ice affects geomorphology and open-water conditions as well as geochemical and biological conditions (Prowse 2001a: 1). The river ice regime consists of river freezing, the main winter period, and ice breakup. In the breakup periods, the natural dynamics of the river ice can be observed, in which the ice most often careens downstream at speeds in excess of 5 m/s and increases water levels by greater than 1 m/min (Prowse 2001a: 8). Ice-jam flooding is a typical extreme event within this type of phenomenon. The role of such floods in supplying organic material and nutrients to floodplain areas, however, is reasonably well documented (Prowse 2001b: 27).

A case of ice-jam flooding in Canada was examined from the perspective of climate change research. In a study of the Peace-Athapasca Delta, it was shown that global warming has resulted in a 2- to 4-week decrease of the ice regime period, thinner ice cover, decrease in the frequency of ice-jam floods, and a decline of biological diversity (Beltaos et al. 2006). The case demonstrates ways in which global climate change can affect a regional ecosystem.

Nonetheless, previous research has not addressed the effects of such changes to regional ecosystems on human society. Although Arctic and subarctic regions are characterized by low population density, river basins have been a typical target for human land use. This will likely continue into the future, so it is necessary to examine the way in which spring floods, which in these basins typically include ice-jam flooding, affect local human societies.

The purpose of the present study is to examine a critical issue, namely, relationships between spring flooding and local human societies in the era of climate change. This is done with a focus on the Lena River in Siberia.

Ice-jam floods occur every year in the Lena River, primarily because it runs south to north (Prowse 2007: 206). To be precise, the river originates from the southern Baikal region and discharges into the Arctic Ocean. It has the longest distance of river ice cover in the world (Bennett and Prowse 2010). The amount of freshwater discharge into the Arctic Ocean is so great so that it affects not only the hydrological and thermal conditions of the arctic region but also the global climate (Ma et al. 2000). Therefore, the Lena is a perfect case for studying the effect of spring flooding on human society as a human dimension investigation of global climate change. The study also describes the concrete effects of ice-jam flooding on the local population that has historically lived along the northern river banks.

The present study incorporates both concrete details of the thawing process in the Lena River basin, explaining characteristics of ice movement and flooding, and an evaluation of the impact of this process on local society.

To this end, it is necessary to follow a holistic approach in the study of the arctic system that incorporates the river and human inhabitants of the river basin. The thawing process, movement of drifting ice, and ice-jam flooding must be described concretely and then connected to perceptions of the local population of the thawing regime and spring flooding. In this way, the research synthesizes multidisciplinary scientific analyses.

7.2 Review and Purpose of Study

7.2.1 Catastrophic Flooding and Climate Change, Lena River

This research is based upon there having recently been an increase in spring flooding of the Lena River (Filippova 2010). Specifically, the frequency of major floods has increased in the past 200 years. There were six spring floods in the nineteenth century and eight in the twentieth century connected to the Lena River that were reported as causing damage to the city of Yakutsk (Filippova 2010). In the first decade of the twenty-first century, there have already been seven catastrophic floods in Yakutsk (PRS 2010).

According to a Russian government resolution (PRS 2010), the Sakha Republic has been cited as the region in the Russian Federation that has suffered the greatest number of extreme flood events in the last 12 years. Because of climate change on both a global and regional level, devastating floods occur much more frequently in this region than in other regions and causes great economic loss. During periods of catastrophic flooding, the amount of damage has reached as much as 21% of the republic's annual expenditure (2001). When there is less disastrous flooding, economic losses range from 1.2% to 1.7% of that expenditure (Table 7.1).

Generally, the main reason that global warming affects less severe spring flooding lies in the temperature difference with latitude; warmer temperatures at higher latitudes than lower ones change the north–south temperature gradient that modifies the conditions of ice breakup and flooding (Prowse 2007: 206). However, from other perspectives, the frequency of flooding may be identified as the result of agents other than climate change. The Russian government and academics remain undecided as to the reason for the increase of spring floods on the Lena River, seeing the causes as either climate change or anthropogenic factors (Kustatov et al. 2012; Sukhoborov 2006). A government report from 2010 drew attention to human risks from the increased frequency of disastrous spring flooding in this region. Although the decisive cause of these events has not been clarified, the government has urgently demanded adequate policies and measures to deal with the increasing regularity and severity of these events.

To explore the primary reason for this increase, we believe that it is necessary to present data related to the spring flooding process in both scientific detail and human dimensions. Because the spring river ice process and flooding have definite effects on local human societies, these are now becoming urgent issues that are subjects of research by social scientists and disaster management policymakers. From the human perspective, river ice in this region is an important element of the regional infrastructure because it is heavily used as winter roads for transportation in winter (Okumura et al. 2011; Vuglinsky et al. 2006) and for local subsistence activities such as a source of fresh water (Takakura 2012).

The purpose of this work was to clarify the conditions involved in the process of spring flooding on the Lena River. We describe concrete processes such as ice movement and speed, loci of flooding outbreaks, and the effects on local human societies. Description of the thaw regime process including spring flooding was a central task. Understanding the process of river ice breakup and consequent flooding is important for interdisciplinary Arctic studies, because these phenomena are closely related to local societal–technological characteristics and the regional environment.

Table 7.1 Economic loss by flood in Sakha Republic

	1998	1999	2000	2001	2002	2003	2004	2005	2006	2007	2008
Expenditure (mil. Rub)		16,860	24,323	33,348	37,042	41,557	46,342	64,514	60,737	66,096	83,348
Flood Damage (mil. Rub)	939.4	0	0	7000	114.6	0	439	97.4	7.7	1088.5	939.1
%		0.0%	0.0%	21.0%	0.3%	0.0%	0.9%	0.2%	0.0%	1.6%	1.1%

Sources: Regiony (2004) and Regiony (2010)

7.2.2 Trends of Lena River Flooding

To overview trends of the spring flooding, we gathered datasets from an open database[1] and have shown the results in Table 7.2. During the 1998–2008 period, there were eight major flood events, for which the average duration, damage, number of displaced persons, affected area, and severity are listed. During the period, the most disastrous flooding was in 2001; this flood represents the extreme end of the scale in regard to economic loss, number of displaced people, and affected area.

For the floods in the above period, we calculated the correlation coefficient of flood scale using the affected area (in km^2) and linked it with other factors (Table 7.2). There is a positive correlation between the flood-affected territory and economic loss, flood severity, and number of displaced people. It is clearly seen that the larger the scale of the flooding, the greater the economic loss, damage, and number of displaced people. A late starting date of flooding has a weak negative correlation with flood scale. If in fact climate change postpones ice decay, ice-related flooding would also be delayed. This dataset shows the delay of the flood is not necessarily correlated to the scale of the flood.

The most important finding is that there is almost no correlation between the scale and duration of flooding. The −0.05 value of correlation shows that independent of flood duration, large-scale flooding may or may not occur. In other words, either a very short and sudden flood season or prolonged flood period can result in catastrophic flooding under certain conditions.

Thus, the timing and duration of flooding is not strongly correlated to its scale, which points to the difficulties for both scientific understanding of these events and adaptation policies for predicting spring flooding as a natural phenomenon. Therefore, we must focus on each case of spring flooding individually. Here, this is first done through analysis of a local newspaper.

7.3 Methods

The present analysis involves an interpretative approach using multidisciplinary data from satellite imagery, literature covering affected areas, and anthropological fieldwork. Specifically, Landsat or large-scale resolution images of flooding from a bird's-eye view provided a general overview. Information on ice monitoring and flood reports from local newspapers was visualized using GIS software (Google Earth), and Advanced Land Observing Satellite (ALOS) or high-resolution imagery of a particular flooding event were interpreted based on information from our anthropological fieldwork. Through these combined means, we identified several types of flooding and examined their implications for the local human societies.

[1]The Dartmouth flood observatory global archive of major flood events is available at http://floodobservatory.colorado.edu/index.html

Table 7.2 The information of Lena spring flooding from 1998–2008

	Economic loss (mil rub)	Main cause	Periods	Duration	Lateness of starting flood (from May 01)	Displaced	Affected sq. km	Severity
1998	939.4	Snowmelt	May 16–June 06	22	15	51,000	314,800	1.5
2001	7000	Snowmelt	May 12–May 27	16	11	70,000	2,857,000	2
2002	114.6	Snowmelt	May 24–June 01	9	23	1500	85,630	1
2003	0	Snowmelt	May 27–June 20	25	26	400	62,480	1
2004	439	Rain and snowmelt	June 12–June 23	12	42	500	136,400	1
2006	7.7	Rain and snowmelt	June 04–June 12	9	34	1300	274,800	1
2007	1088.5	Snowmelt	May 13–June 13	32	12	7000	313,200	1
2008	939.1	Ice jam-break-up	May 12–May 25	14	11	700	595,300	1

Sources: Dartmouth Flood Observatory Global Archive of Large Flood Events (http://www.dartmouth.edu/~floods/index.html), PRS 2010

Correlation coefficient of flood affected space with other factors

	Duration	**Lateness of start**	**Displaced**	**Economic loss**	**Severity**
Affected sq. km	−0.059	−0.469	0.786	0.990	0.880

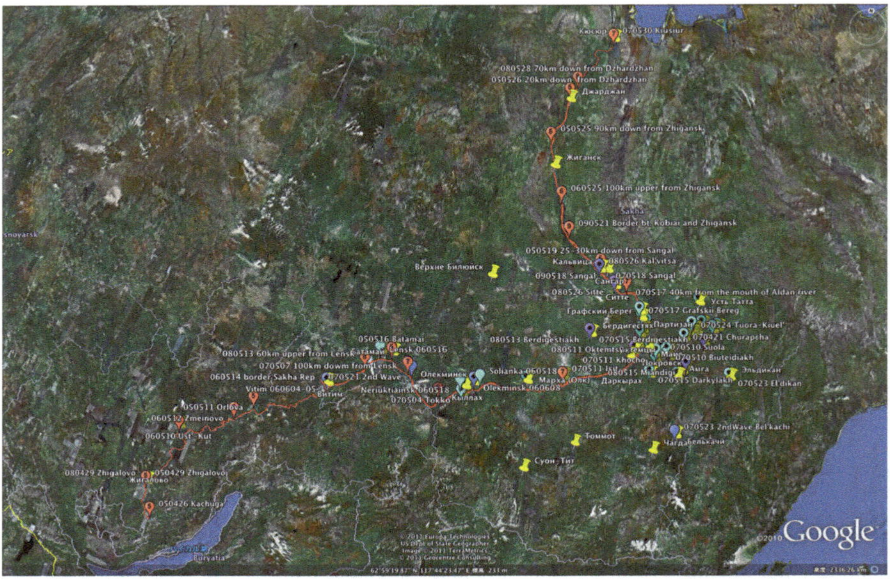

Notes: Red mark with digit means the head of drifting ice with the last digit of year; Blue and purple marks with dot means the disastrously flooded village; Six digit means yymmdd, for exaple, 060515 = 15 May 2006.

Fig. 7.1 Drifting ice movement and flooded village on the Lena River during 2005–2009

The primary data were collected from a newspaper analysis covering 2005–2009. The local newspaper *Iakutiia* features a specialized series entitled "Flood" during the spring flooding season from the last half of April to first half of June. We selected information on certain flooded villages and flood periods as well as location and time information detailing the lower edge of drifting ice, which can describe the movement of those ice floes.

Based on what we collected from the newspaper, we marked information on the flooded villages and locations of drifting ice in Google Earth (Fig. 7.1). We then measured the distance between each of these points, and from this we calculated the speed of drifting ice. We also described the situation by adding related content from the news articles. Through this method, a typology of spring flooding based on causes of the events can be discerned. Information pertaining to flood damage from a particular village combined with locations of drifting ice provided a concrete visualization of the flood. Finally, we combined findings from anthropological fieldwork and interviews with the aforementioned data analysis to illustrate the disaster in the local context.

7.4 Overview of Newspaper Analysis

Table 7.3 shows the number of articles collected from the local newspapers that were related to spring floods. The primary purpose of these articles is to convey daily announcements on the risk of flooding to the capital city of Yakutsk. Therefore, regardless of the actual occurrence of spring floods, each year has a number of these articles. Relatively large-scale floods were reported in 2007 and 2008, and these years had many articles on the flooding.

The contents of these articles were classified as containing six types of information: (a) Flood adaptation policies by local authorities; (b) movements of drifting ice, in particular its lower edge (*nizhniaia kromka ledokhoda*); (c) details on river surface events; (d) water level, (e) heavy flooding reports; and (f) numbers of victims and amounts of socioeconomic loss. Therefore, we could gather information on the location and dates of the lower edge of drifting ice, from which we could identify the movement pattern of these ice floes.

Although the contents of the articles are structured, they are of somewhat different quality, so it is difficult to do a systematic analysis. It is possible to note the arrival dates of the lower edge of drifting ice at Yakutsk, as well as the number of damaged villages and districts (Table 7.4). However, it is more useful to treat the information as case studies or to refer to location data of the drifting ice.

In a previous study, we described information collected from the newspaper articles and then created a list of drifting ice movements and flooding events from 2005 to 2009 (Takakura 2013). To visually comprehend the collected information from a bird's eye view and assemble it for multi-disciplinary analysis, we input all the information on locations of drifting ice in the Lena River and data of flooded villages and districts into Google Earth (Fig. 7.1). In this way, we were able to visually represent the relationship between the flooding disaster and the location of the head of ice drifts.

Table 7.3 Number of flooding article in the local newspaper

2005	10
2006	7
2007	28
2008	17
2009	12
TTL	74

Table 7.4 Changes of the spring flood information

	2005	2006	2007	2008	2009
The arrival date of the head of ice drifting to Yakutsk	18 May	21 May	14 May	19 May	15 May
Number of suffered village	0	3	33	14	4
Suffered district	0	2	9	8	2

7.5 Movement of Drifting Ice: Visualization and Analysis

Table 7.5 shows drifting ice movement calculated from the newspaper information. First, we input the date and place name corresponding to the lower edge location of the drifting ice in Google earth. The number of registered locations ranges from five

Table 7.5 Drifting ice movement 2005–2009 calculated from the newspaper information

#	Date	Place	Latitude	Longitude	Distance (km)	Day	Per day speed	Per hour speed
2005-1	4/26/2005	Kachuga	53°57′46.91″N	105°52′45.95″E	0	0		
2005-2	4/29/2005	Zhigalovo	54°48′55.06″N	105° 9′9.55″E	147	3	49.00	2.0
2005-3	5/11/2005	Orlova	58°17′5.03″N	109° 5′24.10″E	692	12	57.67	2.4
2005-4	5/16/2005	Batamai	60°45′39.58″N	115°36′5.05″E	562	5	112.40	4.7
2005-5	5/19/2005	25–30 km d. Sangal	64°5′35.80″N	127° 5′24.17″E	1258	3	419.33	17.5
2005-6	5/25/2005	90 km d. Zhigansk	67°34′32.04″N	122°59′58.59″E	464	6	77.33	3.2
2005-7	5/26/2005	20 km d. Dzhardzhan	68°56′24.69″N	124° 3′32.40″E	170	1	170.00	7.1
	ttl-average				*3293*	*30*	*147.62*	*6.2*
2006-1	5/10/2006	Ust'-Kut	56°47′52.87″N	105°45′56.06″E	0	0		
2006-2	5/12/2006	Zmeinovo	57°46′50.07″N	108°17′12.49″E	306	2	153.00	6.4
2006-3	5/14/2006	Border Irksk-Sakha	59°22′9.68″N	112°32′15.19″E	394	2	197.00	8.2
2006-4	5/15/2006	Lensk	60°43′22.23″N	114°56′30.12″E	217	2	108.50	4.5
2006-5	5/16/2006	Niuia	60°31′46.44″N	116°15′56.96″E	84	1	84.00	3.5
	5/17/2006	Niuia			–	–	–	–
2006-6	5/18/2006	Del'ge	59°54′22.14″N	118°36′10.21″E	188	1	188.00	7.8
2006-7	5/21/2006	Yakutsk	62°2′34.35″N	129°45′14.43″E	715	3	238.33	9.9
2006-8	5/25/2006	100 km u. Zhigansk	65°54′38.75″N	124°10′8.45″E	592	4	148.00	6.2
	ttl-average				*2496*	*15*	*159.55*	*6.6*

(continued)

Table 7.5 (continued)

#	Date	Place	Latitude	Longitude	Distance (km)	Day	Per day speed	Per hour speed
2007-1	5/7/2007	100 km d. Lensk	60°27′25.48″N	116°39′44.36″E	0	0		
2007-2	5/11/2007	Isit	60°48′55.88″N	125°19′53.23″E	580	4	145.00	6.0
2007-3	5/14/2007	Yakutsk	62°2′34.35″N	129°45′14.43″E	305	3	101.67	4.2
2007-4	5/15/2007	Namtsy	62°43′8.80″N	129° 39′54.15″E	85.3	1	85.30	3.6
2007-5	5/17/2007	40 km d. m. Aldan r.	63°29′19.41″N	128°48′32.09″E	123	2	61.50	2.6
2007-6	5/18/2007	Sangal	63°56′20.18″N	127°28′38.96″E	90.2	1	90.20	3.8
2007-7	5/30/2007	Kiusiur	70°41′3.19″N	127°21′59.43″E	916	12	76.33	3.2
	ttl-average				*2099.5*	*23*	*93.33*	*3.9*
	4/29/2008	Zhigalovo	54°48′55.06″N	105° 9′9.55″E	0	0		
2008-1	5/13/2008	60 km u. from Lensk	60°18′57.39″N	114°16′59.06″E	1188	14	84.86	3.5
2008-2	5/18/2008	Yakutsk	62°2′34.35″N	129°45′14.43″E	1043	5	208.50	8.7
2008-3	5/22/2008	Sangal	63°56′20.18″N	127°28′38.96″E	298	4	74.50	3.1
2008-4	5/28/2008	70 km d from Dzhardzhan	69°17′57.66″N	124°35′44.09″E	693	6	115.50	4.8
	ttl-average				*3222*	*29*	*120.86*	*5.0*
2009-1	4/24/2009	Ust'-kut	56°47′52.87″N	105°45′56.06″E	0	0		
2009-2	5/4/2009	Olekminsk	60°22′36.12″N	120°25′24.79″E	1311	10	131.10	5.5
2009-3	5/12/2009	Tabaga (Y)	61°49′19.59″N	129°37′15.77″E	580	7	82.86	3.5
2009-4	5/15/2009	Yakutsk	62°2′34.35″N	129°45′14.43″E	28.4	3	9.47	0.4
2009-5	5/18/2009	Sangal	63°56′20.18″N	127°28′38.96″E	292	3	97.33	4.1
2009-6	5/21/2009	Border bt Kobiai and Zhigansk	64°56′2.30″N	124°53′42.55″E	175	3	58.33	2.4
					2386.4	*26*	*75.82*	*3.2*
	5 year aver				2699.38	24.6	119.44	4.98

to eight each year, and the total number for the 5-year period is 33. Each location contains date and time information, so we were able to measure the distance between the nearest two points on the map and calculate the speed per day and hour.

The minimum distance that ice was observed drifting in the Lena River is 2099 km (2007) and the maximum 3293 km (2005). Average distance for the 2005–2009 period is 2699 km or 60.4% of the river length (4472 km), and average speeds of drifting ice during these 5 years was 116.8 km per day and 4.9 km per hour. Average speeds (in km/h) in 2005 were 6.2, 6.6 in 2006, 3.9 in 2007, 5.0 in 2008, and 3.2 in 2009.

In hydrologic terms for river engineering, the water movement may be seen as follows. If the water slope ranged from 1/12000 to 1/2000, then the water speed should be 3.20–7.85 km/h. These values were calculated from Manning's formula and, given the assumption conditions, Manning's coefficient of roughness is 0.03 and water depth is 5 m. According to a remote sensing estimate of the Lena River, for a thaw from 13 to 15 May 2007 (Fig. 7.2), the speed of ice movement was 4.2 km/h. From these data, the speed calculated from the local newspaper information is confirmed to be approximately valid.

Correlations of the speed of drifting ice with economic loss and affected area are −0.34 and −0.11, respectively (Table 7.6). Although the sample number is limited, one sees that that increased speed slightly lessened the economic loss and decreased the size of the affected area.

There were 28 observations for speed calculation over 5 years. During this period, the speed of the ice flow was mostly within 1–8 km/h. However, a flow exceeding 9 km/h was found in two cases, 17.5 km/h (16–19 May 2005) and 9.9 km/h (18–21 May 2006).

Fig. 7.2 Landsat images of ice movement in middle basin of the Lena River, 13–15 May 2007

Table 7.6 Economic loss and the velocity of flooding

	Economic loss (mil rub)	Affected sq. km	Velocity
2005	97.4	–	6 2
2006	7.7	274,800	6 1
2007	1088.5	313,200	3 9
2008	939.1	595,300	5
2009	80	–	3 2
Sources	PRS 2010	Dartmouth Flood Observatory Global Archive of Large Flood Events	
Correlation coefficient of ice drift velocity with other factors			
Velocity of drifting ice	−0.34	−0.11	

In the 16–19 May case, according to the newspaper,[2] the 245 people of Dyparai Village in Olekminsk District evacuated because of the risk of flooding during 15–16 May 2006. However, there was no damage and they returned to the village after the departure of drifting ice. In the other case (18–21 May) with the high speed of 9.9 km/h, the newspaper[3] reported that drifting ice movement was observed at Neriuktianinsk and Solianki Villages, 80 km downstream of Del'ge Village. No flood damage was reported along the river from Del'ge to Yakutsk. Interestingly then, severe flooding may not occur with high ice speed.

Speeds were calculated using daily location information, which is usually determined every 2–3 days but sometimes longer. From an observational perspective, drifting ice movement is not constant but rather has stops and starts with variable speeds.

Given the above, a shorter observation duration would increase the accuracy of the calculated speed. There are five cases with calculations over a single day: 7.1 km/h for 25–26 May 2005; 4.5 km/h for 14–15 May 2006; 3.5 km/h for 15–16 May 2006; 7.8 km/h for 17–18 May 2006; 3.6 km/h for 14–15 May 2007. Except for the first, these cases can be examined in more detail because of additional data gathered from newspaper articles treating these events.

According to the aforementioned articles[4] (Fig. 7.3), on the morning of 16 May 2006, an ice jam formed at Niuia Village (84 km downstream of Lensk). Then, the movement of drifting ice ceased and the water level began to increase. The next day, 17 May, the ice jam broke free and began to move downriver to Del'ge Village (Olekma District), arriving on 18 May. The river distance between Niuia and Del'ge is 188 km, so the speed during the motion period was calculated at 7.8 km/h, however there were changes in the speed during these days. According to Table 7.5, there was relatively slow speed from 14 to 16 May, 4.5 and 3.5 km/h on each day. After the ice-jam break on 17 May, the speed of the drifting ice increased more than double, to 7.8 km/h. The ice speed increased immediately after breakup.

[2] The sources are referred from the newspaper *Iakutiia*, 20 May 2005.

[3] *Ibid.*, 19 May 2006.

[4] *Ibid.*, 17, 19, and 20 May 2006.

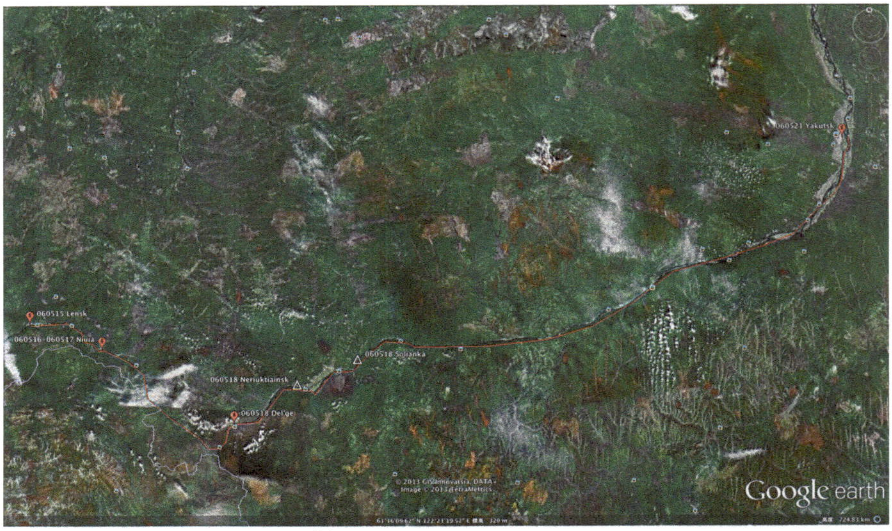

Notes:
Red mark with digit means the head of drifting ice with the last digit of year
Triangle mark means the observation point of drifting ice movement
Six digit means yymmdd, for exaple, 060515 = 15 May 2006

Fig. 7.3 Drifting ice movement, 15–21 May 2006

In a case in 2007[5] (Fig. 7.4), the leading edge of the drifting ice reached Yakutsk on 14 May and then arrived at Namtsy Village at 1 pm on May 15. The distance between these two points on the river is 85.3 km, and the speed of the ice was 3.6 km/h. The floating ice extended from Namtsy to Khangalassi Cape, a length of 53 km. During the day of 14 May, the Markha Villages to the north of Yakutsk were flooded, and this continued through 15 May along the river. Also on that day, a suburb of the city of Yakutsk reported flood damage.

Based the above information, we can confirm that both spring flooding either as water increase in the river channel or severe flooding occurs in conditions with a relatively low speed of drifting ice. The speed of ice drift increases immediately after the collapse of an ice jam. Severe flooding may not occur with a high speed of ice.

7.6 Disastrous Effects of Spring Flooding: Distribution and Types

Figure 7.5 is an image of all villages that suffered flood damage during 2005–2009. We classified four regional groups on the map: the blue circle is the upper Lena Basin and red is the middle basin; the green shows the Amga and Tatta river basins, and the orange is the Alaas region. These are regions frequently subject to spring flooding. The green circle includes branches of the Lena River that flow through the

[5] Sources are from the newspaper *Iakutiia*, 15 and 16 May 2007.

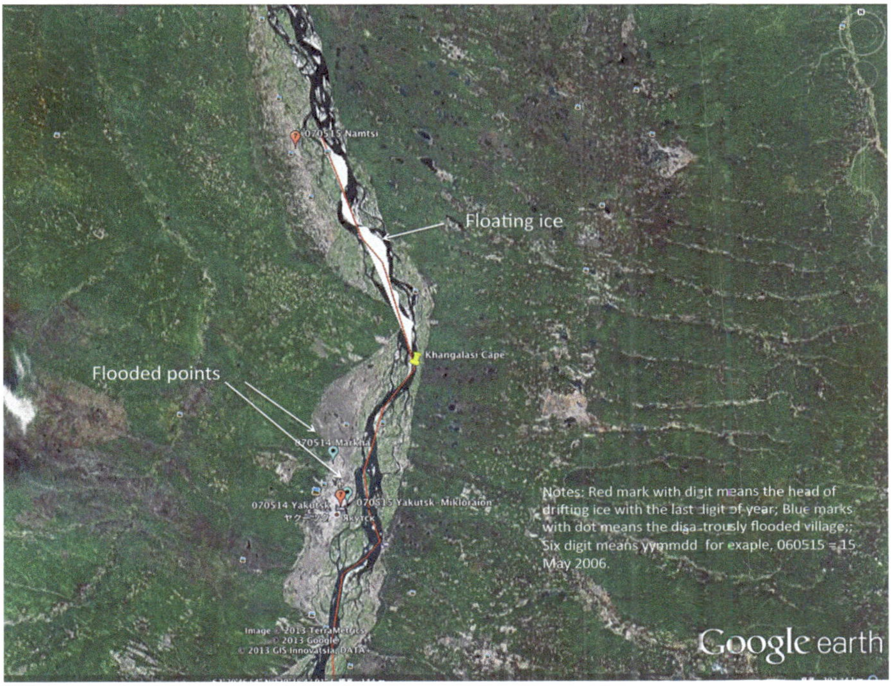

Fig. 7.4 Drifting ice movement, 14–15 May, 2007

Notes: All marks with six digit means the disastrously flooded village; Six digit means yymmdd, for exaple, 060515 = 15 May 2006.

Fig. 7.5 Spring flooding in village, 2005–2009

Notes: All marks with six digit means the disastrously flooded village; Six digit means yymmdd, for exaple, 060515 = 15 May 2006.

Fig. 7.6 Spring flooding in village, 2005, 2006 and 2009

Aldan River.[6] It is interesting that the orange circle contains many open meadowlands with lakes surrounded by forests, and does not include a large river basin.

Figure 7.6 shows villages with flood damage in 2005, 2006, and 2009, when the scale of the spring floods was not large. There were only a few villages that reported flood damage during these years. However, some damage occurred in 2006 in the upper Lena Basin (blue circle).

In contrast, spring flooding was heavy during both 2007 and 2008, for which Fig. 7.7 shows villages sustaining flood damage. The destruction extended over large areas. Villages reporting damage are not only along the Lena River in the red circle, but also in the tributary Amga and Tatta rivers and orange circle (Alaas region).

From the above figures, we see that flooding disasters consistently occur in the Lena River basin, but with larger-scale flooding damage is extensive, both along the tributaries and in the Alaas area. Figure 7.8 shows affected villages along the Lena River during 2005–2009. Field interviews revealed that many residents recognize particular villages along the river in both the upper and middle basin that are consistently flooded in spring.

The green and orange circles in the Fig. 7.7, however, are regions that have only recently experienced more severe spring flooding. Therefore, damage in these areas tends to be serious because the local populations are not accustomed to such events.

When information on an affected village is reported in the newspaper, the cause of flooding is usually mentioned. The reasons are often cited as excessive water

[6] Anna Stammler-Gossman (2012) describes the background and social process of the 2007 disaster in the Tatta region, and discusses the adaptability of the affected village community.

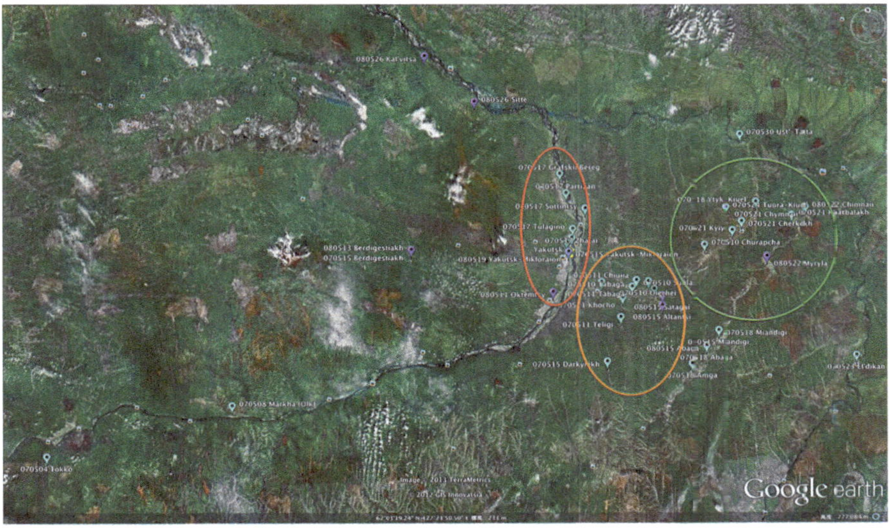

Notes: All marks with six digit means the disastrously flooded village; Six digit means yymmdd, for exaple, 060515 = 15 May 2006.

Fig. 7.7 Spring flooding in village, 2007 and 2008

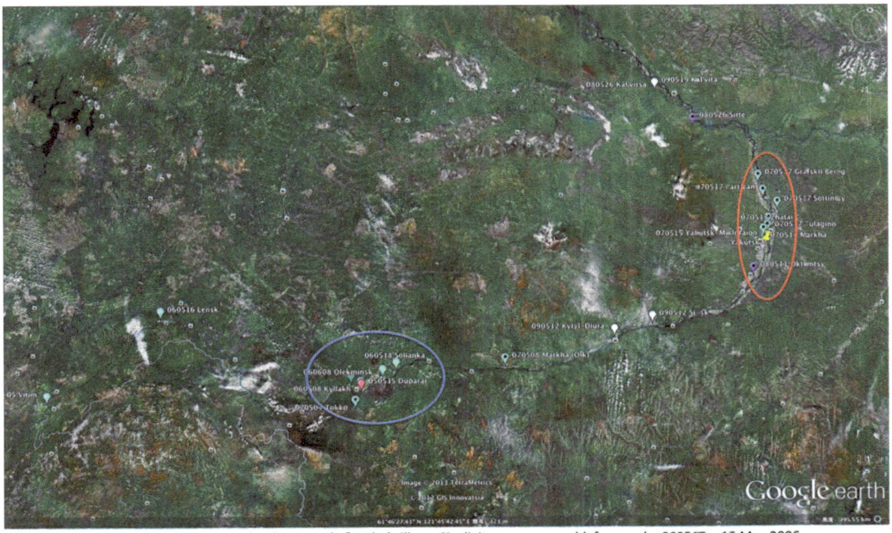

Notes: All marks with six digit means the disastrously flooded village; Six digit means yymmdd, for exaple, 060515 = 15 May 2006.

Fig. 7.8 Spring flooding in village in the Lena River basin, 2005–2009

Table 7.7 The cause of spring flooding of Lena River from the local newspaper

Year	Snowmelt	Measure against flooding	Ice jam	"Second wave"	Flooding in other area
2005	1		1		
2006			2	1	
2007	1	1	3	1	4
2008			3	1	7
2009			2		
TTL	2	1	9	3	11

from snowmelt or water level rise because of ice jams or similar factors. From this information, we discerned five main causes of flooding, i.e., snowmelt, failed measures against flooding, ice jams, secondary flooding after ice-jam floods (locally referred to as "second waves"), and flooding outside the Lena Basin (Table 7.7).

Table 7.7 associates the year with the cause of flooding; from this, one can recognize that ice-jam flooding was regular over the 5-year period. The frequency of these events was similar, at two or three times per year. In 2007 and 2008 when there was large-scale flooding, there were causes other than ice jams in the Lena Basin. Therefore, we must also focus on flooding in other areas, such as in the Alaas region or tributary basins. It can be generally understood that the main cause of spring flooding along the Lena River is ice jams, but the causes of disastrous flooding are much more complex.

7.7 Local Perceptions of Spring Flooding

Finally, we connect these descriptions of spring flooding with anthropological findings to examine how the local population perceives such events. The data are based on fieldwork in Sakha rural villages in the middle Lena Basin, which was conducted from 2008 to 2012. During this time, Takakura, one of the authors was coincidentally present for the large-scale flooding in May 2010.

This event was reported in the federal broadcast of Russia. When the people received warnings from the local administration, they prepared for the flood by moving their valuables up to their attics. Some houses were flooded (Fig. 7.9) and a number of residents in Khangalass district did not prepare, so livestock were killed.

Many informants suggested that spring flooding on the Lena is necessary for pasture management and haymaking preparation. The haymaking activity is seen not only in the alaas, river valley but also in river island (sandbank).

According to one informant, a self-employed cow farmer at Kres-Ktyl Village in the Nam District, "ice jams (*zator* - Russian, *kharyy* - Sakha language) are necessary for the river island. When one forms near the island, the water inevitably

Fig. 7.9 The inside of a house in the Khangalass District, May 2010 (Image provided by the local administration)

appears and then the grass grows well."[7] His meadow for haymaking is on the river island, and his story provides an example of how the ice jam is a natural phenomenon necessary for pasture management in the area's animal husbandry. The river island must be flooded every year, regardless of flood scale. The local population also recognizes the flood as a means of reactivation of the water environment for their subsistence fishing and ice-cube collection for drinking water.

Figure 7.10 shows a flooded-area mosaic image from the ALOS Phased Array type L-band Synthetic Aperture Radar. The first photo shows conditions during summer, in which the green river island is evident. The second photo is a Landsat image taken on 23 May 2007 after the flooding on 14 and 15 May near Yakutsk. It clearly shows that the river island was flooded. Figure 7.11 shows water area estimated from optical and radar satellite sensors, which were 95.6 km^2 (15 July 2010) and 256.8 km^2 (23 May 2007). In spring flooding, the water area increases almost 2.7 times over that in summer.

According to the aforementioned cow farmer, a moderate-scale flood revitalizes the floodplain along the Lena River. This serves to remind us that ice-jam flooding occurred annually during the 5-year period and, unless that flooding was severe, it was seen as beneficial to subsistence activities of the rural population.

[7] Interview with Mr. D. Popov (pseudonym), 17 May 2010.

PALSAR Image		
2007/5/14	Markha village	
2007/5/15	Yakutsk-Mikloraion	
SAR Image	frozen	thaw
forest	violet	green
grassland	dark violet	Light violet
water	violet (white)	black

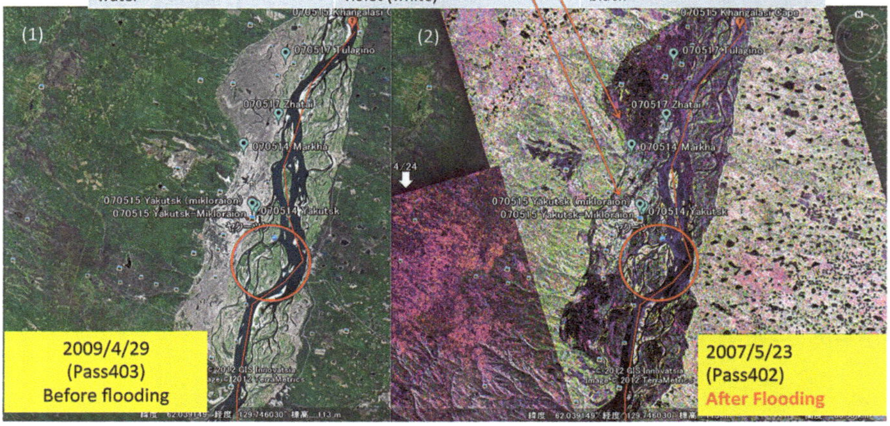

Fig. 7.10 Water area estimated from satellite image of the Lena River, with newspaper information

In a field interview in 2012, a horse herder indicated that the timing and duration of the spring flooding are more important for grass growth than its scale.[8] Even when the flood magnitude was disastrous, its duration on the river island was shorter than at other locations, possibly causing less damage to the population. Another critical element is flood timing. When the flood is delayed for a long period, the growth of grass is severely impaired, which was the case in 2012. Although there was no disastrous flooding that year, the amount and quality of hay was much lower than normally expected. The administration in the region predicted that the number of cattle slaughtered would increase and they prepared to import hay from the Alaas region.[9]

The above information relates a point discussed previously, that the duration of the flood and its delay are not necessarily correlated with its scale. From the perspective of local farmers, an extended duration of water along riversides negatively affects animal husbandry. The disaster policy manager considers this, regardless of flood scale, as an important aspect of ice-jam flooding in human societies. As already stated, the relationship between climate change and increased flood fre-

[8] Interview with Mr. A. Nikolaev (pseudonym), 9 September 2012.

[9] Interview with workers in the local administration at Tulagino Village (14 September 2012), Kres-Ktyl Village (10 September 2012), and Oi Village (18 September 2012).

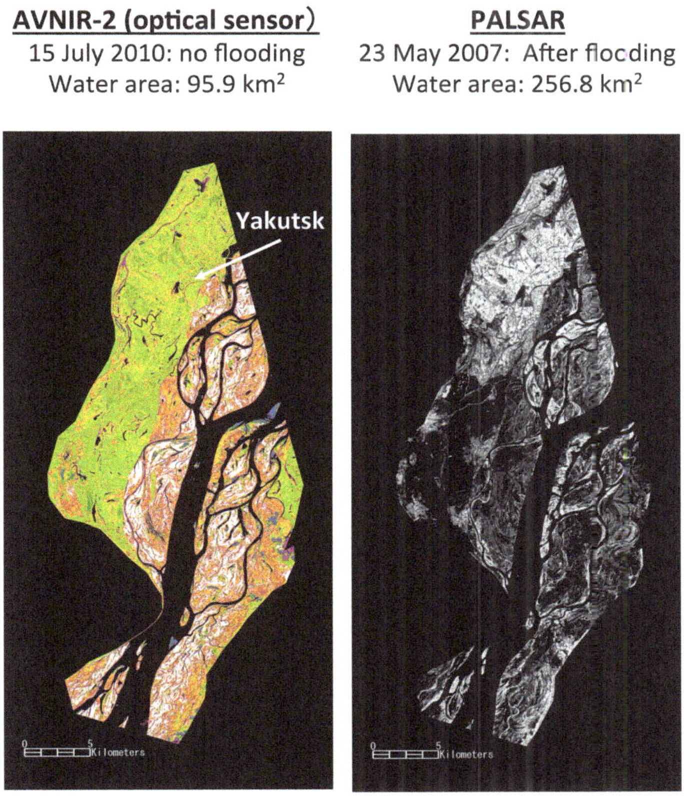

Fig. 7.11 Water area estimated from satellite image of the Lena River

quency has not been confirmed for the Lena River. However, theoretically, the local warming must affect the delay of river thaw and related flooding, and may decrease ice-jam flooding. These two effects definitely have a negative influence on haymaking and local livestock breeding.

7.8 Conclusions

The purpose of this chapter was to explore the process of spring flooding in a northern river basin (the Lena) and its relationship to a local human society. Information from local newspapers, GIS analysis, Landsat, and ALOS images were combined with anthropological interviews. In this way, multidisciplinary data were successfully synthesized. We described the movement of drifting ice, outbreak of flooding, and disastrous events from both a birds-eye view and those of local participants.

With the calculation of drifting ice speed in the Lena River, we examined the ice movement together with various flood factors. The following conclusions were reached.

Flood disaster occurs with relatively slow speeds of drifting ice. The ice speed increased immediately after the ice-jam break. With high ice speeds, catastrophic flooding may or may not occur.

Spring flooding always occurs in the Lena River basin. However, when it becomes disastrous, damage extends beyond the river mainstream. Generally, the main cause of the flooding is ice jams, but with catastrophic floods there is a complex interaction of causative factors.

It is certain that large-scale flooding causes damage to the local population. However, moderate or shorter ice-jam flooding is beneficial to rural agricultural and fishing activities. Therefore, spring flooding is a double-edged sword.

When global warming changes the frequency, scale, or geographic distribution of spring flooding, the effects on human society should be not evaluated simply as either positive or negative. Such change must be understood by incorporating a variety of perspectives and then examined in detail within the local setting. Therefore, an interdisciplinary approach integrating social sciences, natural sciences, and engineering is required for further research into relationships between northern river systems and local human societies, and should be done in association with the study of climate change.

References

Beltaos S (1995) River Ice Jams. Water Resources Publication, Highland Ranch

Beltaos S, Prowse T, Bonsal B, MacKay R, Romolo L, Pietroniro A, Toth B (2006) Climatic effects on ice-jam flooding of the Peace-Athabasca Delta. Hydrol Process 20:4031–4050

Bennett K, Prowse T (2010) Northern hemisphere geography of ice-covered river. Hydrol Process 24:235–240

Filippova V (2010) For the problems on flood in the rivers in Yakutiia during 20th century [in Russian]. In: Gumanitarnye nauki v Iakutii: issledovaniia molodykh uchenykh: Sbor. statei. Nauka, Novosibirsk.

Huntington H, Weller G (2005) An introduction to the Arctic Climate Impact Assessment. In: Symon C et al (eds) Arctic Climate Impact Assessment. Cambridge University Press, New York, pp 1–19

Kustatov KI, Ammosov AP, Kornilova ZG, Shpakovaet RN (2012) Anthropogenic factor of ice jamming and spring breakup flooding on the Lena River. Russ Meteorol Hydrol 37-6:392–396

Ma X, Fukushima Y, Hiyama T, Hashimoto T, Ohata T (2000) A macro-scale hydrological analysis of the Lena River basin. Hydrol Process 14-3:639–651

Okumura M, Kawamoto K, Boyakova S (2011) Winter roads in Sakha Republic and the effects of Global Warming [in Japanese]. Unyuseisaku Kenkyu (Transport Policy Studies, Institute for Transport Policy Studies, Tokyo) 14-3, 16–23

Prowse T (2001a) River-ice ecology I: Hydrologic, geomorphic and water-quality aspects. J Cold Reg Eng 15:1–16

Prowse T (2001b) River-ice ecology II: Biological aspects. J Cold Reg Eng 15:17–33

Prowse T (2007) River and lake ice. In: United Nations Environmental Programe (ed) Global outlook for ice and snow. http://www.unep.org/geo/geo_ice/ [16 Aug 2014]
PRS (2010) Sakha Republic government adopted the Resolution "on the establishment of the concepts of protection for the settlements and economic facilities from the flood" (27 Mar 2010) [in Russian]
Regiony (2004) Reginony Rossii. Osnovnye kharakteristiki sub'ektov Rossiiskoi Federatsii. Ofitsial'noe izdanie. 2004. Stat. Sb. Rosstat, Moscow
Regiony (2010) Reginony Rossii. Osnovnye kharakteristiki sub'ektov Rossiiskoi Federatsii. Ofitsial'noe izdanie. 2010. Stat. Sb. Rosstat, Moscow
Stammler-Gossmann A (2012) The big water of a small river: flood experiences and a community agenda for change. In: Tennberg M (ed) Governing the uncertain: adaptation and climate in Russia and Finland. Springer, New York, pp 55–82
Sukhoborov V (2006) Problems linked to the prevention of natural disasters in the Sakha Republic (Yakutia). In: The conference of "The 3rd Northern forum to the prevention of natural disasters in the Sakha Republic (Yakutia) 2006", http://www.yakutiatoday.com/events/inter_FWG_emercom.shtml [23 Aug 2012]
Takakura H (2012) The local conceptualization of river ice thawing and the spring flood of Lena River under the global warming. In: Proceedings of 1st international conference global warming and the human-nature dimension in Siberia, Research Institute of Humanity and Nature, Kyoto, pp 68–70
Takakura H (2013) The spring flood information of Lena River system seen through the newspaper "Iakutiia" (2005–2009) [in Japanese]. In: Fujiwara J, Hiyama T (eds) FY2012 FR4 Research project report, "Global warming and the human-nature dimension in Siberia social adaptation to the changes of the terrestrial ecosystem, with an emphasis on water environments", (RIHN Project C-07). Research Institute of Humanity and Nature, Kyoto, pp 104–116
Vuglinsky VS et al (2006) Ice regime in the rivers of Russia, its dynamics during last decades and possible future changes. In: Proceedings on the 18th IAHR International symposium on ice. http://web2.clarkson.edu/projects/iahrice/IAHR%202006/contents/12_200.pdf [23 Aug 2012]
Yoshikawa Y, Wanatabe Y, Hayakawa H, Hirai Y (2012) Field observation of a river ice jam in the Shokotsu River in February 2010. In: Proceedings of the 21th IAHR international symposium on Ice, June 2012, pp105–117

Chapter 8
Flood Risk and Migration in the Sakha Republic (Yakutia)

Junko Fujiwara

Abstract Since the late 1990s, the Sakha Republic (eastern Siberia, Russian Federation) has suffered almost yearly flood damage because of climate change. This chapter examines how the local people and government tried to find the most suitable emigration methods to adapt to this situation. The research shows that at first, both the government and inhabitants tried to choose between continuing to live in an old village or abandoning it and moving to a new one. However, after lengthy debate between the government and residents, the government urged not living in places where floods were possible during periods of high risk. This was seen as feasible, because unlike earthquakes or tsunamis, the ice-jam flood risk in the republic occurs only in spring. As a result, the government and residents came upon a beneficial solution, namely, seasonal use of both old and new villages. This type of lifestyle matches the pre-revolutionary traditions of indigenous people and the nomadism currently practiced by some residents of the republic. Therefore, this plan was understood and accepted by the local people. We can say that the Sakha people adapted to flooding with the use local knowledge.

Keywords Climate change • Flood • Risk • Adaptation • Migration • Local knowledge

8.1 Introduction

In the extreme cold of Siberia, rivers freeze in winter and thaw in spring. Traditionally for the people of the Sakha Republic in eastern Siberia, the phenomenon of thawing rivers and flowing ice (*ledokhod* in Russian) has been a sign of the coming spring. Therefore, people looked forward to this time and regarded it as a festive season. However, since the late 1990s, residents' views of the drifting ice have changed

J. Fujiwara (✉)
Kobe City University of Foreign Studies, 9-1 Gakuen Higashimachi, Nishi-ku, Kobe 651-2102, Japan
e-mail: junko@inst.kobe-cufs.ac.jp

T. Hiyama, H. Takakura (eds.), *Global Warming and Human - Nature Dimension in Northern Eurasia*, Global Environmental Studies,
DOI 10.1007/978-981-10-4648-3_8

greatly. A massive ice jam flood occurred in spring 1998, and since then people have suffered almost yearly flood damage. Now, the sight of drifting ice in spring is recognized not as cause for celebration but as a risk. The increased flooding is thought to be associated with global warming, and further flooding is expected in the future (Gosgidromet 2005). This chapter describes how the local people and government are trying to adapt to the situation. Specifically, the research examines various methods of migration and relocation).

Detailed information on flood damage and countermeasures are described in an official document released on May 27, 2010, entitled "Protection Conception of Settled Areas and Facilities for Economic Activities of the Sakha Republic from Flood and Other Negative Influences of Water" (Kontseptsiia 2010). However, there is little information available about the process of migration. This chapter reveals how the migration plan outlined by the republic government is executed. The information here is based on my field research from 2008 to 2011 in the Sakha Republic as well as on newspaper articles and other local sources. The research clarifies processes used by the government and local residents to find suitable relocation methods that would satisfy all stakeholders.

8.2 General View of the Study Region: The Sakha Republic (Yakutia)

The present study examined the Sakha Republic in eastern Siberia in the Russian Federation. Forty percent of the Republic's territory is located above the Arctic Circle. The climate is extremely cold. The northern part of the republic is tundra, and taiga spreads to its south. The mean temperature of the coldest month (January) is −30 to −40 °C in most of the republic. The lowest recorded temperature is −71 °C. The warmest month is July, when the mean temperature in central Sakha ranges from 15 to 19 °C. The ground is covered by permafrost throughout the republic (IKAIa 2007).

The Sakha Republic occupies approximately 20% of the land area of the Russian Federation, but its 2002 population was only 949,000, with 64.5% living in urban areas and 35.5% in rural areas (in 2007) (RSKA 2009). The majority of the population are indigenous Sakha (Yakutian) people (49.9%) and Russians (37.8%), who emigrated to the area after the seventeenth century. Additionally, there are small groups of indigenous people of the northern regions, such as the Evenks (2.2%), Evens (1.6%), Dolgans (0.2%), Yukaghir (0.1%), and Chukchis (0.1%) (As of 2010).[1]

The official languages of the Republic are Russian and Sakha. Also, the languages of minor indigenous populations are spoken. Even in villages where the majority population is northern indigenous people, people often use Russian in everyday life. Thus, minor languages are endangered, and there are many people who do not know their ethnic languages (especially among younger generations).

[1] http://www.gks.ru/free_doc/new_site/perepis2010/perepis_itogi1612.htm (Accessed 23 Jun 2014).

Fig. 8.1 People preparing and airing fish taken from a river (Photo by author)

Consequently, there have been social movements to protect minority languages and cultures.

In rural areas, the Sakha people generally engage in subsistence farming including horse and cow breeding, fishing, hunting, potato and vegetable farming, and berry gathering. The Evens, Evenks and Yukaghir breed reindeer instead of (or in addition to) horses and cows.

Villages in the republic generally feature the following facilities: a government office, public hall, post office, school (elementary to high), kindergarten, hospital or medical office, power station, boiler house, stores, and successor companies of *sovkhoz* (state-owned farms). Because the *sovkhoz* almost completely collapsed following the end of Soviet regime, generally there is little employment in the villages. Currently, schools are often the major employers in rural areas.

The Sakha Republic is rich in aquatic resources. There are more than 700,000 rivers of greater than 10 km in length (Kontseptsiia 2010). Villages are generally along riversides. Some villages are on sandbanks and nearly all are at risk of flooding (Kontseptsiia 2010). Despite this risk, residents choose to live by the riverside for a number of reasons. First, it is convenient for fishing (Fig. 8.1), which is an important protein source for local inhabitants. In the Sakha Republic, people fish year round, and ice fishing is also popular. Another important reason for living near rivers is the presence of sandbars and floodplains, which have rich meadows for oxen and horses to graze (Fig. 8.2).

Generally, villages in the republic have no water service facilities, and this is another reason why people live near the rivers. Drinking water is supplied by water wagons, which draw water from the river. Water for other uses is drawn by bucket

Fig. 8.2 Reaping hay in the rich meadows of a sandbar of the Lena river (Photo by author)

Fig. 8.3 Person carrying water from a river. There are houses along the river (Photo by author)

or pumped up from the river (Fig. 8.3). Therefore, living further from a river means greater inconvenience.

Moreover, residents try to live near rivers because they are important routes for transportation. There are many swamp-like areas in the republic and few villages are accessible by land year round. In many villages, people travel by boat in summer and commute by car on frozen rivers in winter (Fig. 8.4).

Fig. 8.4 A truck travels along the winter road of the frozen Lena River (Photo by author)

8.3 Increased Risk of Freezing from Global Warming

Spring flooding in the Sakha Republic is related to specific factors. As in most colder climates, the rivers freeze over in winter and melt in spring. However, in the republic, rivers flow basically from south to north, so there is a temperature difference between upstream and downstream. In spring, rivers begin to melt in the warm southerly upstream and large quantities of huge ice blocks flow through the rivers. In the downstream, northern parts, there is still thick ice on the river surface. Either ice thickness or sharp bends in the rivers can impede the flow of ice blocks, causing them to clog the river. As a result, the water level rises (Sakai 2015; Hatta 2015).

If the rise of water or spring flood is moderate, it is useful for subsistence. Such floods create fertile meadows, and fill local lakes with fish carried in from the rivers. For the residents, the spring flood has been important for subsistence (Takakura 2013). However, severe flooding may destroy houses and cause other serious damage. Flooding that was previously seen as a blessing is now considered a risk).

The turning point in residents' views of the annual river overflows was the massive spring flood in 1998. This was the first great flood in recent years. A total of 47,000 people were affected in 205 townships. There were five fatalities, more than 15,000 houses were flooded and 746 destroyed completely (Kontseptsiia 2010). The second huge flood in recent years was in 2001. Ten of the 35 districts (including the capital of the republic and surrounding areas) were affected, and there was a small number of fatalities. The city of Lensk and some townships were completely submerged. In total, 3489 houses, 704 agricultural facilities, dozens of bridges, and more than 4000 supply lines for heating, water, electricity, oil, and gas were destroyed. Additionally, 2184 domestic animals died, and nearly 30 tons of agricultural seed were lost (Kontseptsiia 2010). Since that time, as the following Table 8.1

Table 8.1 Damage by flooding in Republic of Sakha (Kontseptsiia 2010)

Year	Amount of damage (rubles)
1998	939,400,000
2001	7,000,000,000
2002	114,600,000
2004	439,000,000
2005	97,400,000
2006	7,700,000
2007	1,088,500,000
2008	939,100,000

shows, the government has been forced to spend large amounts on flood damage almost every year.

According to research by the author at the Ministry of Emergency Planning, in the twentieth century prior to 1998, the water level in rivers was low, so flooding disasters only occurred once every several decades. Currently, according to the above government document "Protection Conception of Settled Areas and Facilities for Economic Activities of the Sakha Republic (Yakutia) from Flood and Other Negative Influences of Water", all townships in the republic are at risk of flood disasters. There are 92 townships with a total population 136,400 where the risk is particularly high (Kontseptsiia 2010). Moreover, about 14% of the population of the republic is exposed to high flood risk.

Given the heavy flood damage in the republic in 1998 and subsequent frequency of extreme flooding, the government began to search for methods to mitigate these disasters. Floods are likely to occur almost every year, bringing serious damage. Therefore, the spring floods have become the primary problem facing the government.

At the governmental level, the Sakha executive office in charge of flood-related disasters and various republic ministries cooperate to organize rescue operations, investigate damages, and address reconstruction and disaster prevention. Local residents also began to take measures aimed at preventing damages as the risk from spring flooding heightened. On an individual level, people evacuate to higher ground and stay in tents, or they evacuate to other villages. When they leave their houses, they preserve important documents, electrical appliances, and furniture by placing them in attics or on roofs, and moving domestic animals to places of safety. At the village level, people measure water rise in rivers daily, keep up-to-date lists of motorboats for evacuation, and move hospital inpatients to safe villages or towns).

Fortunately, there have been relatively few deaths, and recently, fatalities have fallen to near zero. This has been the result of information on the timing and location of river ice thaw being reported in real time, so local inhabitants can predict flooding and evacuate in a timely fashion. However, residents often do not have sufficient time to evacuate domestic animals. In addition, flood damage in which houses are flooded or swept away by rising waters is difficult to prevent, even though shore protection efforts have been made as a preventive measure against flooding.

In the construction process following serious flood damage, the issue of primary concern for victims and the government is whether the reconstruction will be accomplished by winter. Newspapers and other media frequently report on this process. Summer in the far north is short; the peak of summer is July, and by August temperatures start to drop. If snow begins to fall, construction work is disrupted. If alternative housing for flood victims is not provided by winter, they will die from exposure. Even if damaged houses have not been completely swept away, much restoration is needed before winter. For example, flooded houses can become "cold". Owing to wet soil in the basement absorbing heat from the house, these houses do not stay warm enough for people to live in, even if heating systems work. Therefore, the damp soil must be replaced with new dry soil. If firewood for heating is washed away, stocks must be replenished. Whether the residents can accomplish this work within the 6 months or so following the spring flood is a matter of life and death in an area with such extreme cold. The people must work quickly in summer to prepare for the onset of winter. As mentioned above, the increase in size and frequency of flooding is believed to be attributable to global warming. Ironically, this warming has increased the risk of death from exposure to freezing winter temperatures).

8.4 Long Process of Migration

When a village experiences frequent flood damage, the government suggests migration. According to the Sakha Republic executive office for flood relief, if more than 60% of a village territory is damaged more than three times, the population should move to a safer location. However, people do not generally want to leave their home village, even if they experience great hardships.

A comparison with Japanese migration policy is helpful. After the Great East Japan Earthquake and tsunami, the government urged residents to move to higher ground. However, according to Yoshitaka Motoda, professor of traffic engineering at Iwate University, historically in Japan, people migrated to higher ground after tsunami disasters but eventually returned to lowlands. No one desires to be a tsunami victim, but safety is not the only issue when deciding where to live. For those who work in the fishing industry, the most important in coastal areas, it seems reasonable to live near the ocean. Therefore, people tend to return to the coastal lowlands (Motoda 2011).

Similarly, in the Sakha Republic, the loss of daily convenience is the greatest problem in relocating the population. However, this issue has been addressed by paying attention to characteristics of the flooding risks in the republic and by adapting to local traditions, following lengthy debate between the government and inhabitants. Below, I examine how these tasks were achieved based on the examples of 14 villages with flood damage (Fig. 8.5, Table 8.2).

Fig. 8.5 Location of 14 villages

8.4.1 Investigation of 14 Villages

Among the 14 villages, according to a 2002 governmental decision, the inhabitants of 10 villages would be collectively relocated to a site (village Nos.1-10). To understand these 10 examples, I will refer to two villages that ceased to exist after resident migration (village Nos.11-12), and two villages where a migration plan has been proposed but there has been no decision to relocate at present (village Nos.13-14).

Table 8.2 Villages studied

Village	Population(as of 2008)	Flood type	Migration
1. Kutana	600	Spring Ice-Jam flood	Collective
2. Khatystyr	1500		
3. Cheriktei	500		
4. Kytyl-Diura	500		
5. Kal'vitsa	200		
6. Khapchagai	100		
7. Kyllakh	1100		
8. Berezovka	400		
9. 2-i Khomstakh	600		
10. Arbyntsy	300		
11. Bordoi	–		Individual
12. Saldykel'	–		
13. Argakhtakh	600	Permafrost thawing	Not decided
14. Andriushkino	800		

From 2008 to 2011, I conducted field research in two villages (Khatystyr and Argakhtakh), and interviewed affected persons from four other villages that I did not visit (Kutana, Kyllakh, Berezovka, and Andryushkino). In addition, I collected information on all 14 villages from regional newspapers, an official republic website, and other internet references. Furthermore, I interviewed Sakha government executives assigned to flood disaster relief efforts and individuals from the Ministry of Emergency.

The most serious flooding in the republic is that in spring, but I will also refer to villages affected by other flood events to focus on residents' feelings about migration. These other floods are caused by permafrost thaw, which is believed to be caused by global warming. Such floods occur when water from thawing permafrost flows into the river (Gotovtsev et al. 2009).

All 14 villages are near rivers. The following presents a basic overview of these waterways. The Lena is the greatest river in eastern Siberia. It originates in the Baikal Mountains and flows into the Arctic Ocean. It has a length of 4400 km, and the river basin area is 2,490,000 km^2. The Aldan River originates in the Stanovoy Mountains and flows into the Lena. The Aldan is 2273 km long, with a basin area of 729,000 km^2. The Alazeya River flows through the Kolyma lowlands and into the Arctic Ocean. It is 1590 km long and has a basin area of 64,700 km^2.

8.4.2 Method of Migration

There are many possible of methods of migration. The following two examples illustrate how a certain method is chosen.

Example No. 1: Argakhtakh Village

Argakhtakh Village (village No.13) is in the Srednekolymsky District and Alazeia Basin. It has a population of about 600, mainly of Sakha (Yakuts) ethnicity. Following 1997, this village was affected by serious flooding caused by permafrost thaw. This type of flood is considerably different from a typical flood, which is usually of brief duration. Flooding from permafrost thaw continues for a long period. After a spring thaw, the water level gradually rises and this high water level is maintained during the summer. Subsequently, the water freezes at this high level in winter. When the village is flooded, it appears as a small island in an ocean. Because these periods of being surrounded by water are long and frequent, the inhabitants complain about such recent flooding, stating that "… we were on the water for 10 years."

In 2007–2008, when a large flood struck the village, 29 households were damaged and the residents had to live with relatives or acquaintances. During these 2 years there was no air travel between the village and Srednekolymsk, the district center, because the village runway was under water. Grazing land and grasses around the village for oxen and horses were also submerged, causing serious problems for subsistence. Furthermore, in almost all households, underground storehouses were submerged, damaged, or were unable to maintain cooler temperatures over the summer. The village is very remote, so large underground storehouses for refrigeration and freezing are necessities. Currently, people complain that food preservation has become much more difficult because of the flooding. In addition, riverbank erosion accelerated following the more frequent floods. Some facilities prone to submersion were moved to higher ground. However, if the erosion continues to advance, it is expected that the village will become uninhabitable.

When a migration plan was proposed following large-scale flooding in 2007–2008, only 27 residents agreed to relocate to the town of Srednekolymsk. When I inquired as to why people did not want to move, they answered that it is too expensive to live in such a town, because in these more urban areas people must pay for everything in cash. For instance, in the village, residents can catch fish from the river, collect wild berries in the forest, and harvest potatoes and vegetables from the fields. The people never starve in a village as long as they can make the effort to find food.

If they must relocate, the majority of inhabitants would prefer mass migration in which the entire village is relocated. The republic government has searched for a suitable locale for collective migration. However, at the moment this plan is stalled, because this locale has not been found and the water level has fallen since 2009.

Example No. 2: Andryushkino Village

Andryshkino village (village No.14) is in the Nizhnekolymsky District and Alazeia Basin. The population is about 800, and it is designated an ethnic village of the Yukaghir people, one of the minor indigenous peoples of the northern regions. It is a village neighboring Argakhtakh (example No. 1 above), and residents were also affected by flooding from permafrost thaw after 1997. Because it is situated on riverine alluvial soil, the ground of the village is very fragile, so the situation is more serious than in Argakhtakh. In 2007–2008, when a strong flood struck, conditions were such that the entire village was in danger of being swept away.

At that time, various migration plan ideas were proposed. One was to build apartments in Chersky, a town at the center of the district, and have the villagers migrate collectively. The population of Chersky is about 3200, and it is 450 km from Andryshkino. However, the population of Chersky is primarily ethnic Russian, so the republic government and Yukaghir villagers were afraid that the latter's ethnic culture would not survive in the new location. In addition, it was expected that in Chersky they would be unable to continue their subsistence, such as reindeer herding, and it would be difficult to find employment. There were also fears that these conditions would increase alcoholism.

A second idea was to expand a neighboring village and migrate collectively. But in this case, people who worked at schools, government offices and other public facilities in Andryshkino would lose their jobs. In addition, it would be difficult to continue their subsistence, because the new location is distant from the original production sites for breeding, fishing, hunting et al. Therefore, the village assembly rejected the plan.

The best plan appeared to be the construction of a new village near the present one, but currently this option faces obstacles. There is no suitable site close to the village, and the water level in the river has fallen over the last few years.

The above two examples suggest that the most suitable choice for migration is to collectively migrate the entire village to an open site as near as possible to the original location. There are some notable merits to this option. First, it is possible to maintain relationships among villagers. In the republic, most villages have a large number of residents who are relatives, and relations based on mutual aid are important. The continuation of these relationships provides security. Second, in a new village, people can be partially self-sufficient, so they can survive with little income. Third, they can use the same production spots (grazing lands, grasslands for harvesting, fishing and hunting spots) around the previous village, so they can continue their subsistence as before. Fourth, in a new village there would be workplaces in newly constructed public facilities, lessening the danger of unemployment. Finally, this option offers the best chance that the ethnic culture will be maintained. In the republic, not only for Andryshkino but also in other cases, there is an emphasis on protecting the ethnic culture of minor indigenous populations of northern areas. To do this, it is necessary to avoid the collapse of a community and to protect traditional subsistence. Methods that migrate villagers collectively to nearby places are thought to protect ethnic cultures. This type of migration was chosen in all 10 villages where the government decided the population should migrate.

8.4.3 Who Makes the Final Decision, the Government or Residents?

In 2002, the republic government issued an official ordinance[2] to transfer 10 villages seriously affected by flood damage in 1998 and 2001 to higher ground. In this ordinance, higher ground is defined as a location where the probability of submersion by floodwater is less than once every 100 years. Subsequently, in 2003, the government ordered relevant departments to begin moving the villages.[3] However, this decision was not final. There were a number of developments in each case. The following section looks at examples of two villages where emigration has been promoted.

Example No. 3: Kytyl-Diura Village[4]
Kytyl-Diura village (village No.4) is in the Khangalassky District and the middle basin of the Lena River. The population is about 500, which is largely ethnic Sakha. On 4 May 1997, the Lena overflowed and flooded 74 homes housing 296 people. In addition, the power station, kindergarten, bakery, and four boiler houses were flooded. The flooding lasted 10 days. Subsequently, the village sustained spring flood damage almost yearly.

At first, in 1998, the republic government decided to transport the village to an area known as N'urgunu. However, the villagers had no desire to move the entire village and opposed the plan. They wanted to move only some houses that had frequent flood damage to nearby higher ground. This choice was approved in the village assembly, and the district government began to prepare for this option.

However, 63 villagers opposed the plan and submitted a petition to the republic president in April 2002. In this petition, they pointed out that the new location was unsuitable for construction because of underground ice, which might force them to move again to N'urgunu several years later.

In 2009, a village assembly was held again, and the residents decided to transfer the village to N'urgunu, with 83% of the population in agreement.

Example No. 4: Kal'vitsa Village[5]
Kal'vitsa village (village No.5) is in the Kobiaisky District and middle Lena Basin. The population is about 200, and is mainly ethnic Sakha. After 1998, the village was struck by flooding almost every spring. For example, on 21 and 22 May 2008, 86 houses were flooded.

---

[2] Sakha Republican ruling of 15 Jan 2002 No. 22 "About the transfer of most affected by the floods of 1998 and 2001 settlements of Sakha Republic (Yakutia)".

[3] The order of the government of Sakha Republic (Yakutia) in 7 Mar 2007 N 211-p "About the organization transfer of the localities most affected by spring floods in 1998 and 2001".

[4] I referred to the following news sources for Kytyl-Diura: *Yakutia* 24 Sep 2002; *IaSIA* 9 Jul 2009; *Sakha Life* 6 Jul 2009 (http://sakhalife.ru/node/15950 Accessed 8 Sep 2013).

[5] I referred to the following news sources for Kal'vitsa Village: *Respublika Sakha (Yakutia) – ofitsial'nyi server* 2008.6.10; *IA REGNUN* 26 May 2008; *Yakutsk Online* 22 Feb 2011 (http://www.ya-online.ru/news.php?id=297379&page=94 Accessed 3 Oct 2013).

The republic government decided to move the village in 2002. However, the villagers strongly objected to the plan. However, after suffering flood damage almost every year for a decade, in 2008 the village assembly decided to transfer the village, with more than 90% of the people in agreement. Subsequently, a budget was allocated, and construction of a new village commenced in 2011.

From these two examples, we see that the republic government responded to the opinions of the villagers to the extent possible. After the government ordinance to relocate 10 villages, the government waited a long time for the residents to agree. In example No. 3, several stakeholders, such as the republic government, district government, and village assembly, did not agree on migration policy, and the final decision was made only after concurrence of the village assembly. Also, in example No. 4, the republic government respected dissenting opinions of the villagers and waited 6 years for the village assembly to agree to the plan.

Therefore, in most of the 10 cases there was a considerable time lag between the republic government decisions and the start of construction of a new village. There are even villages where, as of 2013, no construction had begun.

8.4.4 Scale of Damage and Thresholds for Continued Village Existence

The extent to which the intentions of villagers are considered appeared to depend on the scale of the disaster. In the two aforementioned cases, the inhabitants experienced damage from water inundation, but their houses were not swept away. In contrast, damage in the following two examples was devastating. Nearly all houses and village facilities were either swept away or pushed off their foundations.

Example No. 5: Saldykel' Village[6]

Saldykel' village (village No.12) was in the Lensky District and Lena Basin. It was completely destroyed by the spring flood of 2001. After the flood, only 17 of 174 houses remained. Electricity lines and telephone communications were also destroyed, and vehicles were unable to pass through the village. The immediate issue was whether the government could secure houses for all the victims by winter. The government let most villagers emigrate to Lensk, in the center of the district. A decision on 26 October 2001 by the republic government stated that Saldykel' village was to be abandoned. This brought grief to residents, because they had lived as a large family.

[6] I referred to the following news for Saldykel': *Yakutia* on 28 Jun 2001, 3 Jul 2001, 3 Aug 2001, 21 Sep 2001, 22 Sep 2001; Sakha Republic ruling of 26 Oct 2001, No. 558 "About exceptions of rural settlement Sardykel' in Lensky District from the credentials of the administrative-territorial division of the Republic of Sakha (Yakutia)".

Example No. 6: Bordoi Village[7]
Bordoi village (village No.11) was in the Tomponsky District on a sandbank in the middle Aldan Basin. It was completely destroyed by the 2001 spring flooding. After the flood, only two heavily damaged, two-story houses remained. Again, the critical issue was whether the government could arrange housing for all the victims by winter. The government decided to abandon the village because the damage was so extreme. All villagers were relocate by the republic government individually to the town of Khandiga, in the center of the district, or to nearby villages.

Compare examples Nos. 1 through 4 with Nos. 5 and 6. From these, we see that whether the villagers' opinions were respected and whether the villages continued their existence depended on the scale of damage. That is, the threshold of continued existence is whether it was possible to secure houses for all victims by winter. When the damage was too severe, a village could not be rebuilt by winter and the government had to find resettlement sites, frequently breaking up the village. Residents had no choice but to follow the policy or risk freezing to death. After being dispersed, it becomes impossible to speak with a unified voice and negotiate with the government and, as a result, the village becomes extinct.

Example No. 7: Kutana Village[8]
Kutana village (village No.1) is in the Aldansky District and Aldan Basin, and has a population of about 200. It is designated an ethnic village of the Evenki people, one of the minor indigenous peoples of the northern regions. In 1998, the village was devastated by spring flooding. At that time, nearly 100 houses were swept away by water. The official republic newspaper reported that the village had essentially ceased to exist.

Under special consideration of the republic government, which attempts to protect ethnic culture, it was decided that a new village be built nearby on higher ground. However, it was impossible complete all houses before winter. Therefore, the government decided to first build four dormitories over the summer. These dormitories would be used as schools, child-care facilities, hospitals, and the like, following completion of the houses. The construction of the new village and migration began within the year and, following this, the village was added to the 2002 migration plan by the government.

Damage to the village in example No. 7 was catastrophic, as it was in Nos. 5 and 6, and it was impossible to finish reconstruction before winter. However, in this instance, the republic government made a special effort to preserve the village, because it was a designated ethnic village for an indigenous minority. This is similar to example No. 2. From these cases we see that for the republic, government protection of ethnic culture is very important.

[7] I referred to the following news sources for Bordoi: *Yakutia* on 30 May 1998, 5 Jun 2001, 3 Jul 2001, 4 Jul 2001, 3 Aug 2001.

[8] I referred to the following news sources for Kutana: *Yakutia* on 6 Jun 1998, 8 Jul 1998, 26 Aug 1998, 25 Sep 1998.

8.4.5 Resistance of Villagers After New Village Construction

In example No. 7, when houses were destroyed in the original village and ones built in a new village, the relocation to the new site went smoothly because there was nowhere to live but in the new village. However, in the case of villages with only inundation damage, the relocation process was much different. After the villagers agreed on a policy in a village assembly and construction began, there were still hurdles to overcome until the migration was complete, as seen in the following case.

Example No. 8: Kyllakh Village[9]

Kyllakh village (village No.7) is in the Olekminsky District on a sandbank in the middle Lena Basin. The population is about 1100, mainly ethnic Sakha. After the end of the 1990s, the village was flooded in spring almost yearly. At first, people strongly objected to migration. This was because the village was very beautiful, with rich farming and grazing lands and splendid meadows. However, house foundations rotted and the repeated flooding damaged buildings such as schools, kindergartens, and hospitals. Repairs became financially fruitless, because the village was flooded every spring. People were exhausted with the situation, and the village assembly unanimously approved to move the village in 2002. Construction of a new village began in 2006 in a place known as Dapparai, three kilometers distant from the old village. However, the media seemed pessimistic as to whether people would actually live there. The following is an article from the official republic newspaper *Yakutia,* dated 21 September 2007.

> Many residents of Kyllakh are skeptical about the resettlement to Dapparai. First, this is because there is little room to graze cattle and make hay. There is little land suitable for agriculture. So, housing built in Dapparai will only serve as an evacuation point.

According to my interview with a government official, the residents were not going to move even after houses were built in the new village. The old village is nearer the river, so it is convenient for fishing and other everyday activities. Additionally, the land is more suitable for cultivation of potatoes and other vegetables, and there are rich meadowlands. If the people did not live in the new village on which the government had expended a large budget, this would cause a great deal of trouble for the government. Therefore, the government made desperate efforts to persuade the villagers to move. They basically said that the villagers could reside temporarily in the old village during summer if desired, but they had to move to the new village. Finally, the villagers began to relocate.

The government logic is that the people can move according to the season. In other words, residents can live in the new village from winter through the end of the spring flood and in the old village during summer if desired. What are the merits of this option?

---

[9] I referred to the following news sources for Kyllakh: *Yakutia* on 18 Apr 2002, 21 Sep 2007, 23 Sep 2008.

First, people are safe during the spring floods in the new village, because it is on higher ground. Second, the government can save money that would have been spent repairing flood damage. The most important thing for the republic government is to reduce annual reconstruction costs. Compensation for individuals is paid only when their main residence is damaged. So, if people have registered a new house in the new village as their main residence, the government does not have to pay compensation for damage to a house in the old village. Thus, where the villagers choose to live in summer is not important for the government.

Third, people can continue subsistence hunting, fishing, and agricultural activity as in the past. For the northern population the short summer is the most important season for food production to prepare for the winter. In summer, people are very engaged in activities to this end, such as cultivating potatoes and other vegetables, gathering berries, reaping hay for horses and cows, preparing preserved foods. If they can stay in the old village during the summer, which is convenient for such subsistence activity, they can produce food as usual and store it for winter.

Finally, warm housing for the winter is guaranteed. There is a risk that houses in the old village will suffer flood damage. However, if these houses are only for summer use, it is not necessary to be concerned with insulation. Even if a house is washed away, simple summer houses can be easily rebuilt. There is no problem if the foundations of these houses get wet from flooding, causing the heating systems to not function effectively. In short, flooding will have no effect on warm housing for residents in winter.

In fact, changing living space seasonally is traditional for indigenous people in the republic (IKEA 2007). Before the domiciliation policy implemented in the early Soviet era, it was common for the Sakha or other minor indigenous peoples to change locations to facilitate subsistence. In addition, in the Soviet Era, sovkhoz working parties seasonally lived far from their villages. Even now, some people such as reindeer herdsmen, fishermen, and hunters still follow this lifestyle. Given this cultural background, the suggestion by the government to live in the old village in summer and a new village in winter was very understandable and accepted by the local people. In this way, the first complete relocation among the 10 villages and collective migration was realized in Kyllakh.

The new lifestyle in Kyllakh, living in two villages seasonally, was introduced in the village of Kytyl-Diura to persuade people to accept a migration plan. Subsequently, the village assembly agreed to relocation. Such a lifestyle is also evident in Khatystyr village (Fig. 8.6).

8.4.6 Extended Relocation Process

One of the characteristics of migration in the republic is that it is carried out over the long term. Houses and public facilities such as schools, kindergartens, government offices, power stations, and boiler houses are gradually moved from old to new villages. It is extremely expensive to relocate a village, and the republic government

Fig. 8.6 Khatystyr: progress on construction of a new village (Photo by author)

cannot fund this process in one bulk payment. The government has to move the facilities gradually, and it is beneficial if the old and new villages are near each other. Even if public facilities extend over two villages, for example, the kindergarten and government office remain in the old village and the school in the new village, this is not a major problem but may be inconvenient. Some people are engaged in construction of houses in the new village during daytime, and go home to the old village at night.

As important facilities move to the new village, residents' motives to relocate increase. Specifically, moving the schools attracts people to the new village. Usually, there is a school bus (in Kyllakh, a special village ship for pupil during the summer) between the old and new villages during the migration, but this is a hardship for children waiting for the bus in winter. So, young families want to move along with the school. In addition, schools have been the largest employers following the collapse of the state farms in the Soviet era. Teachers and other workers at schools are usually positive about relocation. The years that relocated schools began to operate in villages (where I could confirm the start of construction of new villages) are shown in Table 8.3.

8.5 Conclusions

This chapter has presented how the republic government and local residents negotiated and found optimum migration methods. Below is a brief summary of the process and the migration methods.

Table 8.3 Opening of relocated schools in new villages

Village	Year
1. Kutana	2005
2. Khatystyr	2007
5. Kal'vitsa	Not yet (as of 2013)
7. Kyllakh	2010
8. Berezovka	2010
9. 2-i Khomstakh	2012

People prefer collective migration when migration is unavoidable, and the republic government attempts to choose this method whenever possible. However, the threshold of existence for a village depends on whether it is possible to secure houses for all persons affected prior to winter. When damage is too severe, the government does not attempt to rebuild a village and people must follow the relocation policy or face freezing to death over the winter months.

At first, both the government and inhabitants tried to choose between continuing to live in an old village or abandoning it and moving to a new one. However, after lengthy debate between the government and residents, the government urged not living in places where floods were possible during periods of high risk. This was seen as feasible, because unlike earthquakes or tsunamis, the ice-jam flood risk in the republic occurs only in spring. As a result, the government and residents came upon a beneficial solution, namely, seasonal use of both old and new villages. This type of lifestyle matches the pre-revolutionary traditions of indigenous people and the nomadism currently practiced by some residents of the republic. Therefore, this plan was understood and accepted by the local people. We can say that the Sakha people adapted to flooding with the use local knowledge.

The above solution satisfied almost all stakeholders. People began to spend the spring flood period in safety without completely deserting beloved hometowns. The convenience of subsistence farming in summer was not lost. Cultures of minor indigenous peoples were not subject to potential extinction. The government was able to reduce the budget for flood damage compensation. Thanks to the close locations of old and new villages, people could adapt to the slow migration process, with budgetary savings. Rather than relocating the village, this process is actually an expansion of the existing village. People have adapted to the flood risk by expanding the village area and moving its nucleus.

In the Sakha Republic, people have adapted in this way to the increased flood risk since the late 1990s. The desire in northern territories to meet winter with plentiful food stores and warm housing was satisfied. It seems that in the republic, this style of migration will be adopted in future, wherever possible. However, this type of plan is not applicable to all villages. For example, it is impossible when there is no suitable area for migration nearby. In addition, this method does not apply to places subject to longer periods of flooding, such as the permafrost flood in the Alazeia Basin. In such cases, it will be necessary to search for new methods that can be adapted in the future.

References

Gosgidromet (2005) Federal'naia sluzhba po gidroteorologii i monitoring okruzhaiushchei sredy (Gosgidromet). Strategicheskii prognos izmenenii klimata Rossiiskoi federatsii naperiod do 2010–2015 gg. i ikh vliianiia na otrosli ekonomiki Rossii. Moscow. http://geo.tsu.ru/content/faculty/structure/chair/meteorology/library/pdf/StrProg.pdf. Accessed Oct 2010

Gotovtsev SP et al (2009) Alazeia -2008: Razrabotka nauchno-obosnovannykh rekomendatsii po zashchite naselennykh punktov v basseine r. Alazei ot negativnogo vozdeistviia vod (Goskontrakt No.017-2009). Yakutsk

Hatta S (2015) Siberiano kasenryushutu. In: Hiyama T, Fujiwara J (eds) Siberia – water and social environments in the warming far North. Kyoto University Press, Kyoto

IKAIa (2007) Istoriko-kul'turnyi atlas Iakutiia: priroda, istoriia, etnografiia, sovremennost'. Feriia, Moscow

Kontseptsiia (2010) Kontseptsiia zashchity naselennykh punktov i ob"ektov ekonomiki Respubliki Sakha (Iakutiia) ot navodnenii i drugikh vidov negativnogo vozdeistviia vod. http://www.sakha.gov.ru/docs/253.pdf. Accessed June 2010

Motoda Y (2011) "Kyosei subekidenai takadai iten", Traffic engineering: 46-5. http://p-www.iwate-pu.ac.jp/~motoda/koutuukougakuteigen.pdf. Accessed 14 Mar 2014

RSKA (2009) Respublika Sakha (Iakutiia): Kompleksnyi atlas. FGUP «Iakutskoe aerogeodezicheskoe predpriiatie»

Sakai T (2015) Hanrangen no nobokuchi riyouto kikohendo. In: Hiyama T, Fujiwara J (eds) Siberia – water and social environments in the warming far North. Kyoto University Press, Kyoto

Takakura H (2013) "Is an ice-jam flood a disaster? Indigenous knowledge of river ice and adaptation of the Sakha communities in the middle basin of the Lena River", Northeast Asian Studies: 17, Center for North Asian Studies

Chapter 9
Reindeer Herding and Environmental Change in Reindeer Herding Regions of the Sakha Republic: Comparison with the Yamal-Nenets Autonomous District

Atsushi Yoshida

Abstract I investigated the influence of global warming or climate change on the indigenous peoples of Siberia, especially reindeer herders. Historically, despite problematic management systems (governmental, municipal, or private), these indigenous peoples have adapted to various natural and social changes. This chapter reports on field research into nomadic summer and winter camps in reindeer herding regions (specifically, the villages of Olenek and Sebyan-Kyuelj) of the Sakha Republic, with comparison to West-Siberian Yamal-Nenets reindeer herders. This was accomplished by direct interviews with herders and local administrative and enterprise executives. Some informants told of direct or indirect influences of meteorological change, but most were unaware of change in vegetation for reindeer forage or noted little influence of climate changes. I detected recent vegetation change around the herding camps for pasture conditions by examining normalized difference vegetation index (NDVI) images. Using this method even over short periods, one could detect certain anomalies of vegetation productivity. However, it was difficult to discern the meaning of such anomalies and verify their cause and effect. In future research into the influence of climatic change on the indigenous peoples of Siberia, we must consider its natural and social implications.

Keywords Reindeer herding • Climate change • Sakha • Yamal-Nenets • NDVI • Vulnerability

---

A. Yoshida (✉)
Faculty of Letters, Chiba University, 1-33 Yayoi-cho, Inage-ku, Chiba 263-8522, Japan
e-mail: ayoshida@faculty.chiba-u.jp

T. Hiyama, H. Takakura (eds.), *Global Warming and Human - Nature Dimension in Northern Eurasia*, Global Environmental Studies,
DOI 10.1007/978-981-10-4648-3_9

9.1 Introduction

Russia has 1,583,000 domesticated reindeer, including about 194,900 (12% of all head in Russia) in the Sakha Republic and 683,300 (43%) in the Yamal-Nenets Autonomous District (at the beginning of 2012)[1]. In Russia, there are 17 indigenous peoples of Siberia and the North who have been engaged in this traditional mode of subsistence, from Saame in the west of the continent and Chukchi in the east.

Reindeer herding in Russia has always been under pressure from political and socioeconomic changes and natural resource exploitation, and now global climate change through the twentieth century and into the twenty-first century. Moreover, reindeer herding has been affected by pasture conditions, which are dependent on the herding and husbandry method. In this context, agricultural collectivization is another important factor that has greatly affected reindeer herding, herders, and their societies.

The influence of climatic change on Siberian populations, including indigenous peoples, is not a new topic. Here, I briefly analyze the effects of vegetation and other environmental changes, including the normalized difference vegetation index (NDVI), on herding pastures in two reindeer regions of the Sakha Republic, specifically the villages of Sebyan-Kyuelj and Olenek.

9.2 Potential Influence of Climate Change on Reindeer Herding

In the twenty-first century there have already been several international projects dealing with far northern Eurasian indigenous peoples and their responses to the climate change. These projects have reported various results, including prospects for reindeer husbandry in the research regions. However, there has been insufficient research into East Siberian reindeer herding regions, including Sakha Republic.

Before introducing my field research in the Sakha Republic, I summarize what factors may influence reindeer herding, especially pasture conditions, based on previous research.

- **Anthropogenic changes:**
  - Natural (hydrocarbon, mineral, forest) resource exploitation
  - Industrial expansion and worker influx
  - Building infrastructure
  - Radioactivity (in certain locales)
- **Influence of reindeer herding**
  - Overgrazing and trampling of pasture vegetation
  - Fertilization of pastures from animal feces and urine

[1] Industry Program of Ministry of Agriculture of the Russian Federation Moscow 2013.

- **Climate change**
  - Warming temperature
  - Change in snow–freezing regime
  - Change in freezing–thawing periods and frequency
  - Change in precipitation amount and frequency

In addition, other phenomena that can be observed in various reindeer herding practices in Russia may be related to one or more of the aforementioned factors, such as

- Increase in reindeer predators (wolf, bear, and wolverine)
- Frequent insect harassment (e.g., mosquito and horsefly)
- Decline in reindeer populations

9.3 Field Research: Meteorological Data and Perception of Climate Change Among Reindeer Herders of Sakha Republic and Yamal-Nenets

In the framework of RIHN "Global Warming and the Human-Nature Dimension in Siberia," I performed field research in two reindeer herding enterprises: GUP Sebyan based in Sebyan-Kyuelj village in the Kobyai District (2009) and MUP Oleneksky based in Olenek village in the Olenek district (summer 2010 and winter 2013). For comparison, I will introduce individual reindeer herding practices among Tundra-Nenets in the Yamal-Nenets Autonomous District of West Siberia, where I did field research from 1995 to 2008.

9.3.1 GUP Sebyan

Sebyan-Kyuelj is an isolated village in the Verkhoyansk ridge system of the north-eastern Kobyai District. This village has 800 residents, about 85% of whom are ethnically the Even people. There is a reindeer herding governmental enterprise called GUP Sebyan. It has 16,000 domesticated reindeer (including 5000 private reindeer) that are separated into 10 brigades (with multiple herds). I stayed at one summer camp of brigade No. 9, in the Sulanichan River valley (65°04′N, 129°53′E), about 1000 meters above sea level.

The grazing range of this brigade is small compared with other brigades. Vegetation around the summer camp at the end of August consists of sparse larch on hillsides and surface vegetation, mainly graminoid grass and lichen with mush-rooms. The people of the brigade stay at a winter log house in the southern part of the grazing zone from December through March. In this camp and Sebyan-Kyuelj, I interviewed some Even residents and herders and collected the following information about climate change and perceptible phenomena as follows:

- **Climate situation**
  - "Below average temperature in spring"

    This does not correspond to JAMSTEC meteorological data offered by Mr. K. Yamamoto (Nagoya University) and K. Oshima (RHIN). According to their data, the spring (March–May) mean temperature from 1950 to 2008 has increased (Fig. 9.1).
  - Repeated freeze-thaw cycles affect infant mortality of reindeer calves through the formation of ice crust over snow; this prevents reindeer, especially infants, from reaching forage under the snow cover.
- **River flooding and erosion** – Induced by heavy summer rains, the frequency and extremity of flooding has increased over the last 20–30 years; there is pasture degradation around the floodplain and riverbank erosion. Summer mean precipitation data show wide periodic amplitudes of annual temperature. Residents may perceive every positive anomaly as heavy rain.
- **Change of fauna**
  - Increase of wolves, bears, and wolverines, predators of domesticated reindeer. In mountainous taiga/tundra landscape, these animals attack more than one thousand domesticated reindeer every year. However, relationships between climate change and animal biomass, living place, and migration pattern are not clear.

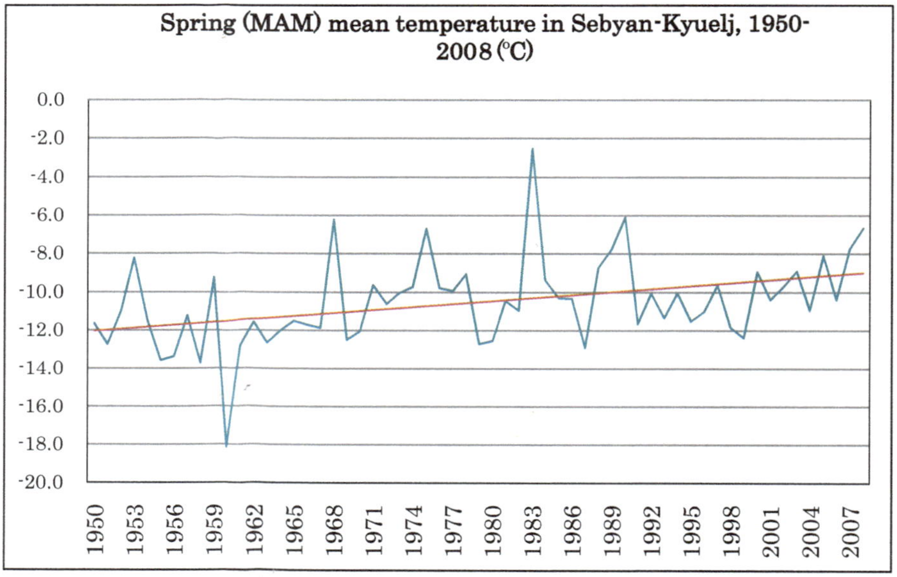

Fig. 9.1 Spring (MAM) mean temperature in Sebyan-Kyuelj, 1950–2008 (°C) (JAMSTEC BMDS by K.Yamamoto (Nagoya Univ.) & K.Oshima (JAMSTEC))

9.3.2 MUP Oleneksky

MUP Oleneksky is in the Olenek District of the northwestern Sakha Republic. Olenek is a central village on the left bank of Olenek River (68°30′15″N, 112°26′50″E). The main office of MUP Oleneksky is in Kharyyarakh village, on the right bank of Olenek River just across from Olenek village. The office was formerly Sovkhoz Oleneksky in the Soviet period. There were 3500 reindeer at the beginning of 2010, separated into three herds for each brigade. (For perspective, the greatest number of reindeer belonging to this enterprise was 34,000, in the 1980s). Reindeer graze during summer far from the village to the northwest, beyond the Sakha Republic and Krasnoyarsk Territory border. There is a treeline in the middle of the grazing range. Its southern part is mainly covered with larch forest, and the northern part is a mountainous tundra zone.

During my first visit, I stayed at six camps of three brigades (Nos. 1, 5 and 6). There, I performed field research in the northern part of the pastures, about 600–800 m above sea level, from the end of August to beginning of September 2010. I stayed at a camp 10 km north of the treeline (69°47′30″N, 109°28′58″E), although there was a sparse larch forest.

My second visit in winter camp in March 2013 was just during the time of gathering the brigade reindeer into the corral where the animals are counted and injections of anthrax vaccine are given. The corrals have been constructed near winter camps, where each herd stays for several weeks until the corral campaign or *koralizatsiya* in early spring season every year.

The following is the result of herder interviews in the camps regarding the environmental situation.

- Climate situation
  - Temperature: *low temperature in summer and high temperature in winter* over the last 20–30 years (Figs. 9.2–9.4).

    Some herders stated that "In summer, high-temperature spells used to last a week or so, after that they weakened. But last time there was no such continuous heat spell.

    Using a short period of meteorological data, one can detect the same winter phenomena as indicated by the herders' perception (Fig. 9.3).
  - Severe winter low temperature in 2012–2013. Such phenomena have occurred every 15–20 years or more during the last 50–60 years. However, winter (December–February) mean temperature from 1964 to 2013 indicates a slight rise (Fig. 9.4).
  - **Permanent snow cover** – Its area is diminishing. However, the influence of such a phenomenon is not clear.
  - **Precipitation** – Relatively heavy rainfall has increased in summer.

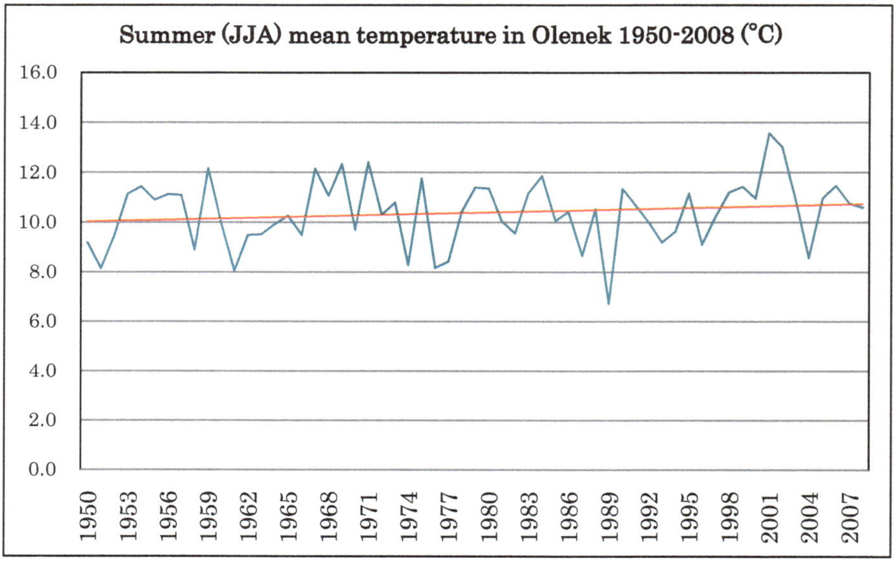

Fig. 9.2 Summer (JJA) mean temperature in Olenek 1950–2008 (°C) (Same resource as Fig. 9.1)

- **Change of flora**
  - A northward and upward shift of the treeline. In the summer pasture of MUP Oleneksky in northwestern Olenek District, the open larch forest is shifting northward.
  - When I asked one of the herders "What bothers you in this situation"?, he answered half-jokingly "It will take longer to pass through the forest zone, which means that it will take more time for us to get to the winter log house."
- **Change of fauna**
  - Wild reindeer migration routes changed in 1976 and since then, there have been frequent conflicts with domesticated herds. In autumn 2010, wild reindeer "abducted" 500 of 1200 domesticated reindeer in a herd of brigade No. 6; of those, only 200 head returned in winter
  - The wolf population increase – In recent winters, this has become a serious problem for reindeer herding. It was discovered that in the Sakha Republic, there are an estimated 4000 wolves, and there has been a yearly increase by 500. There has been a concomitant increase in reindeer kills by wolves. From the beginning of 2013 through April, wolves killed more than 2600 reindeer. In winter 2012–2013, the Ministry of Agriculture of the republic offered a 20,000 ruble reward (700 U.S. Dollars) for killing a wolf.[2]

[2] Information from Internet site http://www.ykt.ru/, accessed 2013/3/14 and 2013/4/23.

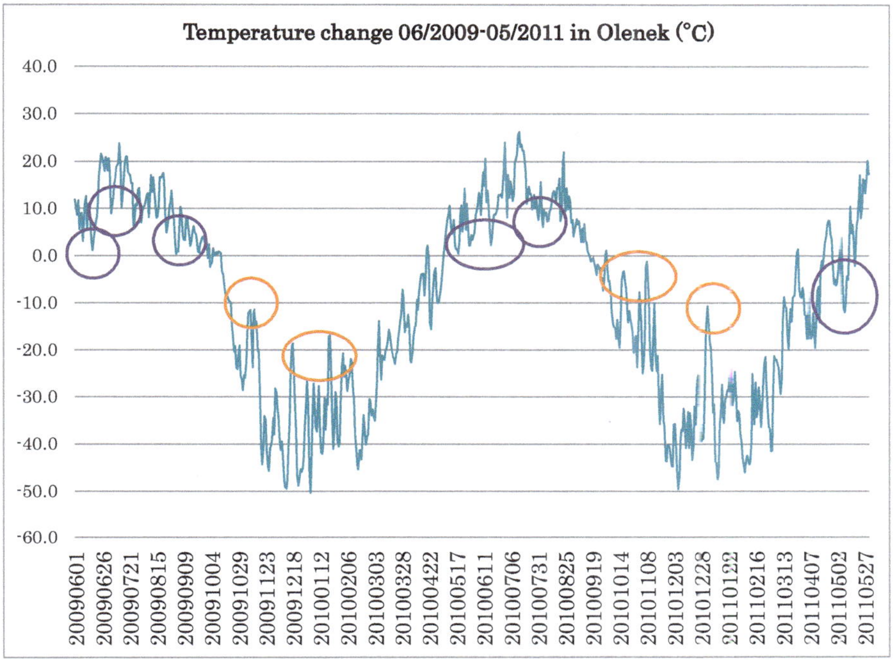

Fig. 9.3 Temperature change 06/2009–05/2011 in Olenek (°C) (Same resource as Fig. 9.1)

9.3.3 Tundra Nenets in Yamal-Nenets Autonomous District, West Siberia

For comparison, I now describe the West Siberian reindeer pasture condition and Tundra Nenets practice, based on published materials. In northern parts of the Yamal-Nenets Autonomous District (further: A.D.), "arctic tundra" or "dwarf shrub tundra" suitable for reindeer grazing are prevalent. A prominent feature of the District reindeer herding is that 53% of domesticated reindeer are privately owned. In contrast, in the Sakha Republic, this proportion is just 10%. In my study area (Tazovsky Region in Yamal-Nenets A.D.), the figure is about 80%. This contrast between Yamal-Nenets A.D. and the Sakha Republic is because of differences between methods of herding, husbandry, land use, and herding management. I will not give a detailed explanation of these aspects here.

There have been several research projects in the twenty-first century in this A.D. According to some of these works, pasture conditions are changing from comparatively stable to a more dynamic regime, as follows.

- **Temperature** – Summer air temperatures have increased around 2 °C over the past 25–30 years (Forbes and Stammler 2009).

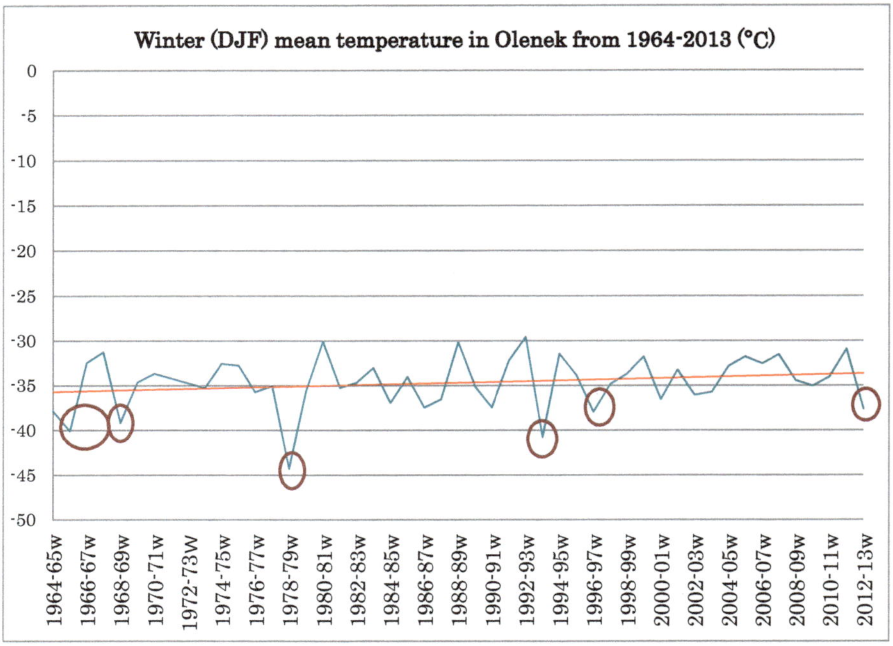

Fig. 9.4 Winter (DJF) mean temperature in Olenek from 1964 to 2013 (°C) (NOAA Satellite and Information Service)

- **Change of flora**
  - There has been a northern shift of vegetation (shrub) (Goetz et al. 2011).
  - There had been a transformation of shrub- to grass- and sedge-dominated tundra where intensive reindeer grazing is practiced (Forbes et al. 2009; Kumpula et al. 2011).
  - De-lichenification has occurred (Bulgakova 2010).
- **Change of fauna**
  - Populations of arctic fox, migratory birds (geese, ducks, swans), sea mammals, and fish (whitefish, family *Coregonidae*) have changed.

Except for arctic fox, which are hunted for fur, the above game animals and fowl are important local foods, especially for nomadic populations. The corresponding change of biomass might affect the food intake balance and thereby the dietetic situation of local populations.

9.4 NDVI Data Comparison of Reindeer Campsites in Sakha Republic and Yamal-Nenets A.D., 2000–2012

It is important to track the change of surface vegetation, which is the primary source of foraging for domesticated reindeer. Such change may be caused by climate change or herding activity (pasture usage). To detect surface vegetation productivity variation as affected by various factors in the aforementioned regions, I have investigated dates of maximum NDVI and presented them in color for visual understanding (Figs. 9.6, 9.7, 9.8, and 9.9). NDVI dates were provided by Dr. K. Yamamoto of Nagoya University.

9.4.1 Sakha Republic: Three Research Areas

The research areas are the same as above, summer and winter camp sites of MUP Oleneksky (see: Figs. 9.5-1 and 9.5-2 – summer and winter camp landscape). For comparison, I also analyzed wild reindeer river-crossing points across the Olenek River.

Fig. 9.5-1 Summer campsite of No.1 Brigade of the MUP "Oleneksky" (2010/8/28)

Fig. 9.5-2 Winter campsite of No.6 Brigade of the MUP "Oleneksky" (2013/2/28) (69°16′28.3″N, 111°16′53.5″E)

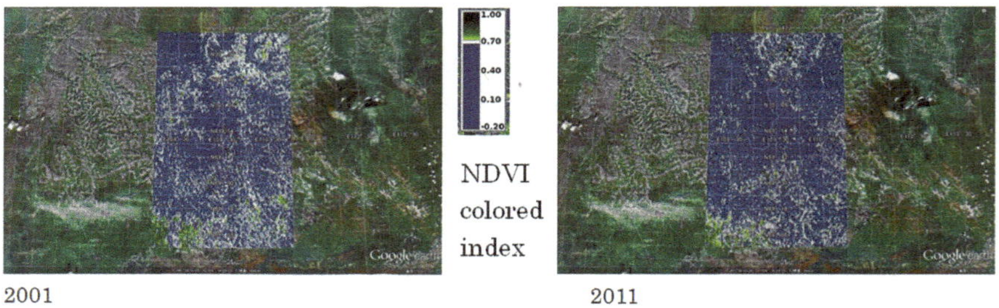

Fig. 9.6 NDVI value images of the pasture area of reindeer herds of the MUP "Oleneksky"

First, I detected change using comparatively small-scale satellite photos, with dimensions 100 km east–west and 150 km north–south. I detected some differences of change over 2000–2012, based on available maximum NDVI values. Two panels in Fig. 9.6 show Google Earth satellite photo insets of maximum NDVI from 2001 and 2011. Comparing the two images, overall it appears that values (green and white colored areas) were large in 2001 and small in 2011. At the image scale, it is difficult to distinguish domestic reindeer herding tracks, migration ranges, or campsite trampling. I therefore used larger-scale images for more detailed information on surface vegetation conditions.

9.4.1.1 Pasture Condition of MUP Oleneksky

Figure 9.7 shows maximum NDVI change at MUP Oleneksky No. 1 brigade's two summer campsites during 2000 to 2011. Image insets at the center of the Google Earth satellite photos indicate three contrasting samples from 2001, 2004, and 2008. There was a minimum in 2004, maximum in 2008 and mean in 2001. Generally, the river basin area had large NDVI values, which are seen in the satellite photo (upper left), but the minimum of 2004 is evident, suggesting some meteorological influence in that year. Based on this analysis, in October 2012, I asked the director of MUP Oleneksky about pasture and weather conditions during 2004. The answer was as follows.

> In autumn 2004, some specialists went to the camp for veterinary and zoological jobs and noticed unusual reindeer fatness. Summer weather that year was cold and rainy and plants did not grow well. The future forecast is for warmer winters and cold and rainy summers ...[3]

By comparing the NDVI information with that from the indigenous population about related areas, we can determine the level of agreement between these two sources regarding vegetable biomass change.

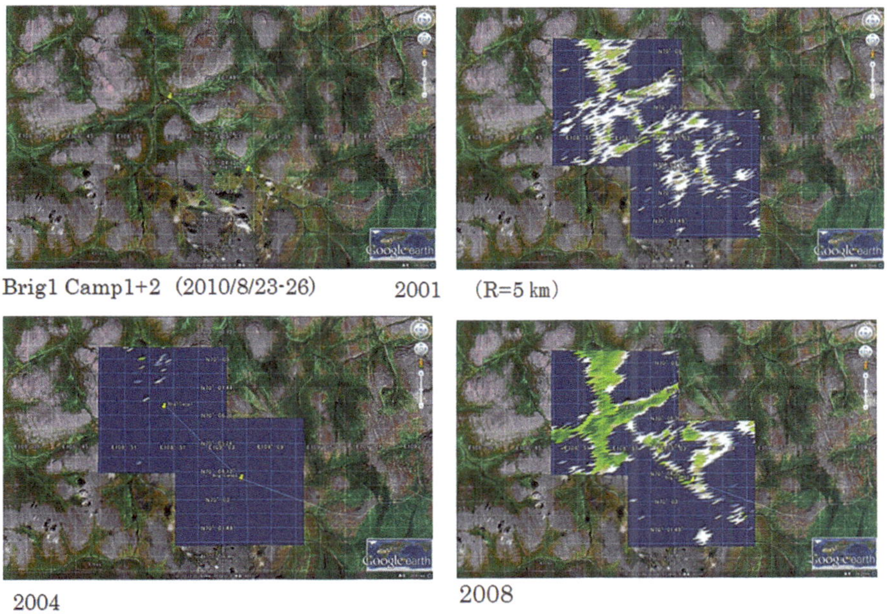

Fig. 9.7 NDVI value images of summer campsites Nos.1 & 2. of Brigades of the MUP "Oleneksky". (Camp No.1: 70°6′51.51″N,108°55′38.076″E; Camp No.2 70°3′47.76″N, 109°5′8.6159″E)

[3] E-mail from director of MUP Oleneksky, M. Kh. Nikolaeva, 29 October 2012.

For comparison, I show another case, that of winter pasture No. 6 brigade campsite of MUP Oleneksky (68°40′10.2″N, 111°40′16.6″E). In this region, larch forest is also prevalent, mixed with rare Siberian spruce (*Picea obovata*). According to some herders, the lichen condition is not poor around the campsite. The maximum NDVI images do not appear to show change over the last decade as compared with northern summer pastures, although values from 2001 and 2004 are smaller than in 2008 (Fig. 9.8).

9.4.1.2 River Crossing Points of Wild Reindeer Across Olenek River

In the latter part of August along the middle of the Olenek River, one can observe river crossings of wild reindeer by a small herd or two or three individuals from the left (northwest) to right (southeast) bank, toward winter pastures. I observed such cases in various places. In the NDVI images, I examined an area around a campsite (68°26′55.00″N, 114°29′45.00″E) for comparison with domesticated reindeer pastures. In comparison with northern tundra and forest tundra regions, values were large. There were no prominent changes detected over 2000–2011.

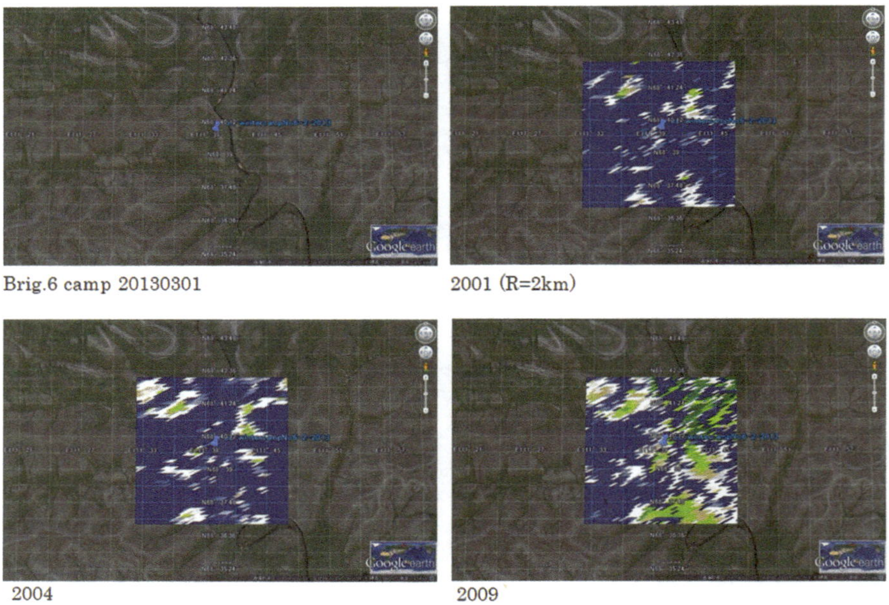

Fig. 9.8 NDVI value images of winter campsite No.6 of Brigade of the MUP "Oleneksky". (Camp No.6: 68°40′10.2″N, 111°40′16.6″E)

9.4.2 GUP Sebyan

Here, I show a set of maximum NDVI images around a summer campsite of GUP Sebyan in the Verkhoyansk Mountains. The images shown in Fig. 9.9 indicate a difference between each image, i.e., small values (prevalent ultramarine area) in 2005 as compared with 2009 or 2011. However, the difference is not as pronounced as MUP Oleneksky's summer pasture NDVI change, described above (Fig. 9.9).

During fieldwork at the aforesaid campsite, herders stated that the most serious problem was pasture erosion by frequent summer flooding and wolf (sometimes bear and wolverine) attacks on reindeer. Such incidents may be caused by climate change or anomalies, which the herders stated have become frequent over the last 20–30 years. These cases might be related to factors other than climate, but this should be analyzed further.

9.4.3 Yamal-Nenets A.D.

There have been research projects including NDVI analyses for Yamal Tundra Nenets. I summarize these based on one such work involving NDVI on the Yamal Peninsula in the western part of the A.D. The NDVI increased with warmer summer

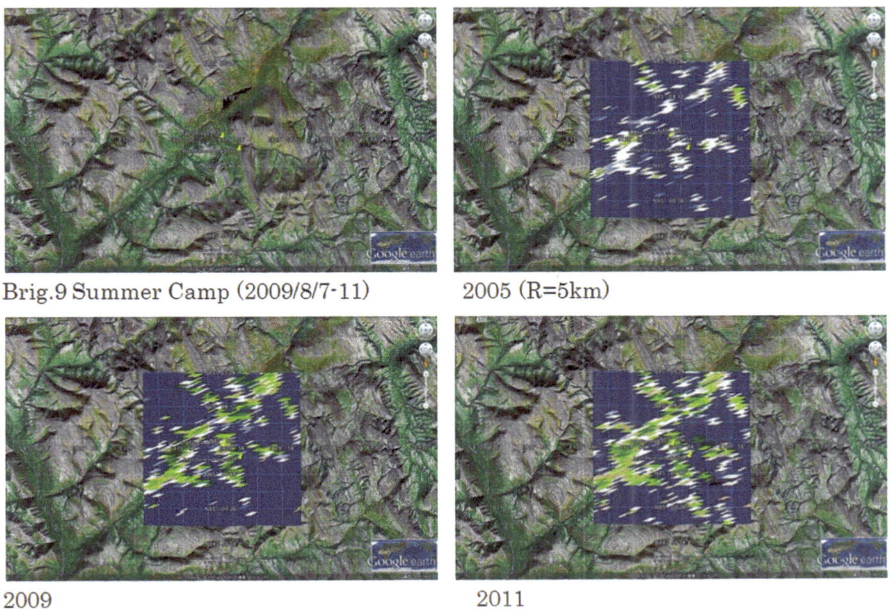

Fig. 9.9 NDVI value images in several years around summer campsite No.9 of Brigade of the GUP "Sebyan", Kobyai District. (65°02′44″N, 129°53′′E)

land-surface temperatures, but this relationship is not as strong as in the Arctic as a whole (Worker et al. 2011). I also examined NDVI imagery from 2000 to 2012 around pastures of the Gydan Peninsula, in the eastern A.D. There, I did fieldwork among Tundra Nenets reindeer herders from 1995 to 2005. However, like the examples of Sakha Republic pastures, the determination of situations surrounding anomalous surface vegetation change was uncertain. Therefore, the images used for analysis of Yamal-Nenets A.D. are not shown.

9.5 Analysis

Generally, the three reindeer regions in the Sakha Republic and Yamal-Nenets A.D. have both common and differing characteristics in the context of environmental change. Each region has its own unique features and problems, which depend on their ecological environment, tundra, mountainous tundra, mountainous taiga, socioeconomic situation, historical background, and ethnic or cultural aspects. One can distinguish the common aspects. From comparing natural and socioeconomic aspects of environmental change among reindeer herding in the regions, several features emerge, which are shown in Table 9.1.

Sakha reindeer herding areas, such as Sebyan-Kyuelj (main ethnic group Even) and Olenek (Evenki) villages, will be socioeconomically vulnerable in the near future. This is because every governmental or municipal enterprise in the regions continue with unprofitable management, even with republic and federal subsidies.

Table 9.1 Natural and Socio-economic aspects of environmental change for the reindeer heeders' communities in Sakha Republic and Yamal-Nenets Autonomous District

	Natural – meteorological/ biological – aspects	Socio-economic aspects
Common aspects	Temperature increase	Nomadic lifestyle
	Northward and upward shift of shrub vegetation or tree line	Local political autonomy (Y; So)
	Wild reindeer migration (S and Yg)	Local/Federal administration's support (affirmative action) – subsidy to entrepreneur, medical-hygiene services, boarding school system to nomadic populations
	NDVI anomalies during 2000–2012 years are detected (esp. So)	
Different aspects	Pasture condition – critical (Yy); warning (Yg); stable (S)	Prevailing public/collective (S) or private (Y) management of Reindeer herding
	Overgrazing(Y)	Non-profitable management (So)
	Pasture degradation by flood erosion (Ss)	Commercialization of products (in practice: Yy; experimental: S)
	Predator attack to domesticated reindeer (S)	Pasture degradation by oil and natural gas exploration (Y)
		Ethnic identity: strong and active (Y); weak and partly assimilating (So)

(Y Yamal-Nenets A.D., Yy Yamal Peninsula, Yg Gydan Peninsula, S Sakha Republic, Ss Sebyan-Kyuelj village, So Olenek village)

Furthermore, the situation of wild reindeer migration has become critical for reindeer herding management in recent years. The problem of animal predation on domesticated animals has also become critical. Especially, in 2011 in the Sakha republic, there were about 13,000 reindeer kills by wolves. These phenomena are somewhat inevitable because the routes and periods of wild reindeer migration and predator habitats are unpredictable, and there are no effective methods of avoidance. In contrast, present pasture conditions do not appear critical, except for erosion caused by summer floods in mountainous taiga and tundra areas (mainly in eastern Sakha). Research in another reindeer herding region of Sakha (Tompo District) showed a similar situation of reindeer herder perception and pasture conditions (Nakada 2012).

In the Yamal (Tundra) Nenets case, the situation appears critical for pasture conditions, given the current resource exploration and industrial development. This situation is more serious in the western part of the region (Yamal Peninsula) and less serious in the east (Gydan Peninsula), though in the latter, a natural gas exploration project is in progress. Fortunately, the strong ethnicity of the Nenets people may help maintain their traditional subsistence system based on traditional ecological knowledge (TEK) or local knowledge (LK).

Maximum NDVI image analysis showed that certain vegetation changes or anomalies have occurred in a relatively short period (2000–2012). However, it is unclear as to whether these phenomena are a direct result of global climate change or regional or periodic anomalies. One can only say that there have been some anomalies in this extreme northern environment, which affect subsistence activities of indigenous peoples in Siberia.

9.6 Conclusions: Nomadic Reindeer Peoples' Vulnerability Under Climate Change

Pastoralism and nomadic lifestyles are highly resilient means of subsistence. These permit nomadic peoples to move not only for their domestic animals but also to avoid natural or environmental difficulties, or occasional social conflicts. Of course, the mobile lifestyle of reindeer herding must be based on the existence of vast regular pastures. In this context, it is important to investigate and analyze pasture conditions and their use from various aspects, including TEK or LK. This will be done in our future research.

In analyzing the vulnerability of nomadic reindeer peoples, it is important to consider not only natural but socioeconomic characteristics. We know that as a type of nomadic pastoralism, reindeer herding is a highly adaptive subsistence system. Moreover, this herding is a very profitable sector of the economy in the far north. A Russian expert indicated that full-year pasture maintenance for reindeer obtained as much as seven rubles of income per one ruble cost of herding

(Podkorytov et al. 2004). This is likely one of the major factors for the existence of reindeer herding throughout the centuries in Northern Eurasia.

We lack sufficient understanding of how societies build adaptive capacity in the face of change (Crate and Nuttal 2009). Therefore, it will be important to examine several approaches to current and future vulnerability studies. To integrate traditional and scientific knowledge, it will be vital to consider and emphasize the social dimensions of climate change (Huntington et al. 2004; Ishii 2011; Stammler-Grossmann 2010).

Acknowledgments I greatly appreciate the assistance of Dr. K. Yamamoto, Nagoya University for providing NDVI images of various regions in Siberia. I also deeply thank the herders and administrative executives of Sebyan-Kyuelj and Olenek villages for advice and support to this research.

References

Bulgakova T (2010) Climate change, vulnerability and adaptation among Nenets reindeer herders. In: Hovelsud GK, Smit B (eds) Community adaptation and vulnerability in Arctic regions. Springer, Oslo, pp 83–105

Crate SA, Nuttal M (2009) Introduction: anthropology and climate change. In: Crate SA, Nuttal M (eds) Anthropology and climate change: from encounters to actions. Left Coast Press, Walnut Creek, pp 9–36

Forbes BC, Stammler F (2009) Arctic climate change discourse: the contrasting policies of research agendas in the West and Russia. Polar Res 28:28–42

Forbes BC et al (2009) High resilience in the Yamal-Nenets social-ecological system, West Siberian Arctic, Russia. PNAS 106(52):22041–22048

Huntington H et al (2004) Matching traditional and scientific observations to detect environmental change: a discussion on Arctic terrestrial ecosystems. Ambio Special Report 13:18–23

Ishii A (2011) A review of research on adaptation and vulnerability. In: FY2010 FR2 Research Project Report (RIHN Project C-07). Kyoto, pp 197–202 (In Japanese)

Kumpula T et al (2011) Land use and land cover change in Arctic Russia: ecological and social implications of industrial development. Glob Environ Change. doi:10.1016/j.gloenvcha.2010.12.010

Nakada A (2012) Reindeer Herding and environmental change in the Tompo district, Sakha Republic. In: Proceeding of 1st international conference global warming and the Human-Nature dimension in Siberia, pp 97–99

Podkorytov FM, Zabrodin VA, Borozdin EK, Laishev KA, Vagin AS (2004) Severnoe Olenevodstvo. Agrarnaya Rossiya, Moscow

Stammler-Grossmann A (2010) 'Translating' vulnerability at the community level: case study from the Russian North. In: Hovelsud GK, Smit B (eds) Community adaptation and vulnerability in Arctic regions. Springer, Dordrecht et al., pp 131–162

Chapter 10
Reindeer Herding and Environmental Change in the Tompo District, Sakha Republic

Atsushi Nakada

Abstract Global warming affects the ecology and behavior of wild reindeer through changes of availability and nutrition of their diet, the distribution of competing species, intensity of insect harassment, and indirectly through accelerating human development activities. The same types of influences on domestic reindeer are also recognized by reindeer herding peoples in many areas of the north. However, although northeast Russia is one of the major reindeer herding areas, these influences have not been investigated. The aim of this chapter is to estimate the influences of global warming on reindeer pastoralism in the Sakha Republic of eastern Siberia. The investigation focused on the Tompo District in the central-eastern Sakha Republic. First, statistical data show that the numbers of domestic reindeer in that district dropped sharply after the Soviet regime collapsed, and that they have gradually recovered in recent years. Second, meteorological data indicate that air temperature has risen gradually in the area, but precipitation has been generally stable. Third, according to my field research, herders managed their reindeer largely in the traditional manner, and they appeared not to clearly recognize climatic change and its influence on the environment and their subsistence activities. These results indicate that although the climate has been changing in Tompo District, it has not harmed reindeer pastoralism. The reasons for this appear to be that the natural environment in the area has not been significantly damaged by climate change yet. Social and/or economic factors may be more important to reindeer pastoralism than climate change.

Keywords Reindeer herding • Environmental change • Even • Sakha Republic • Tompo District

---

A. Nakada (✉)
Hokkaido Museum of Northern Peoples, 309-1 Shiomi, Abashiri-shi, Hokkaido 093-0042, Japan
e-mail: nakada_atsushi@hoppohm.org

T. Hiyama, H. Takakura (eds.), *Global Warming and Human - Nature Dimension in Northern Eurasia*, Global Environmental Studies,
DOI 10.1007/978-981-10-4648-3_10

10.1 Introduction

Global warming is more pronounced at higher northern latitudes, and average arctic temperatures have increased at almost twice the global average rate in the past 100 years (IPCC 2007). Therefore, it is a concern that this warming and related climatic changes have a harmful effect on the health, ecology and population dynamics of reindeer (*Rangifer tarandus*), a species on which the livelihoods of northern indigenous peoples depend (ACIA 2004; Weladji and Holand 2006). Because of their abundance and wide distribution, reindeer have been significant game for indigenous hunters for tens of thousands of years (Gordon 2003; Grayson and Delpech 2005). Reindeer herding instincts and curiosity have made them easy to harvest, and they can be hunted by driving them into V-shaped lanes ending in a corral, net, snares, hidden hunters, lakes or stonewalls (Gordon 2003). Reindeer husbandry probably originated about 3000 years ago, spreading to indigenous peoples throughout northern Eurasia and developing into various forms (Vainshtein 1980; Sasaki 1984). At present, it is the primary livelihood for over 20 indigenous peoples throughout the circumpolar region, and nearly 100,000 people are involved in herding approximately 2.5 million domestic reindeer in nine national states (Oskal et al. 2009).

The general consensus among scientists is that global warming and related climatic variables affect the ecology and behavior of wild reindeer through changes of availability and nutrition of their diet, the distribution of competing species, intensity of insect harassment, and indirectly through accelerating human activities (Klein 1991; Wolfe et al. 2000; Weladji and Forbes 2002; Weladji et al. 2002; Weladji and Holand 2006). Similar influences of global warming on domestic reindeer are recognized by reindeer herders in many areas of the north. Oskal (2008) reported some of these influences that reindeer herders have documented, that is, changes in biodiversity, temperature, precipitation and climate, and indirect effects of climate change. For example, important food resources for reindeer, such as lichens and preferred grass species, may disappear partially, and changes in harassing insect populations could affect reindeer behavior during summer by not allowing them to feed long enough. Change in climate could cause rivers to freeze later in autumn and melt earlier in spring, causing challenges for the annual migration of domestic reindeer. Increasing periods of mild weather with rain followed by cold frost periods form ice layers in the snow and block reindeer access to food on the ground. Furthermore, this may increase accessibility of Arctic regions for human activity that have negative effects on reindeer herds and spoil pasture resources.

Currently, two thirds of the world's domestic reindeer are raised in the Russian Federation (Klokov 2004; Yoshida 2012a). Because of its vast area and various natural, social and ethno-cultural environments, reindeer herding in Russia is diverse and includes several very different economic strategies (Klokov 2012). Some studies dealing with the relationship between reindeer herding and environmental changes, including global warming, have been conducted for both the entire Russian Federation and several localized areas within it (Rees et al. 2003, 2008; Baskin 2005; Forbes and Stammler 2009; Klokov 2012; Yoshida 2012a). However, although reindeer herding in the country has various forms and the influence of global warming appears to vary by locality, there are few studies dealing with this variability, except for western Siberia. Therefore, in this chapter, environmental variables and their influences on reindeer herding are considered for a locale in eastern Siberia, within one of the major reindeer-herding areas of Russia. To approach this problem, three aspects are examined. First, the statistical fluctuation of domestic reindeer number is referred to as an indicator of herding success. Second, fluctuations of climatic variables (air temperature and precipitation) are investigated to verify climatic change. Third, herder awareness of climatic and environmental change and their impacts on reindeer herding are studied to detect sensitivities and responses to such change. These aspects are examined individually, and relationships and interactions among them are discussed.

10.2 Study Area

The investigation focused on the Tompo District in the Sakha Republic (Yakutia) (Fig. 10.1). This republic, situated in eastern Siberia, is the largest federal subject and one of the major reindeer-herding areas in the Russian Federation (Jernsletten

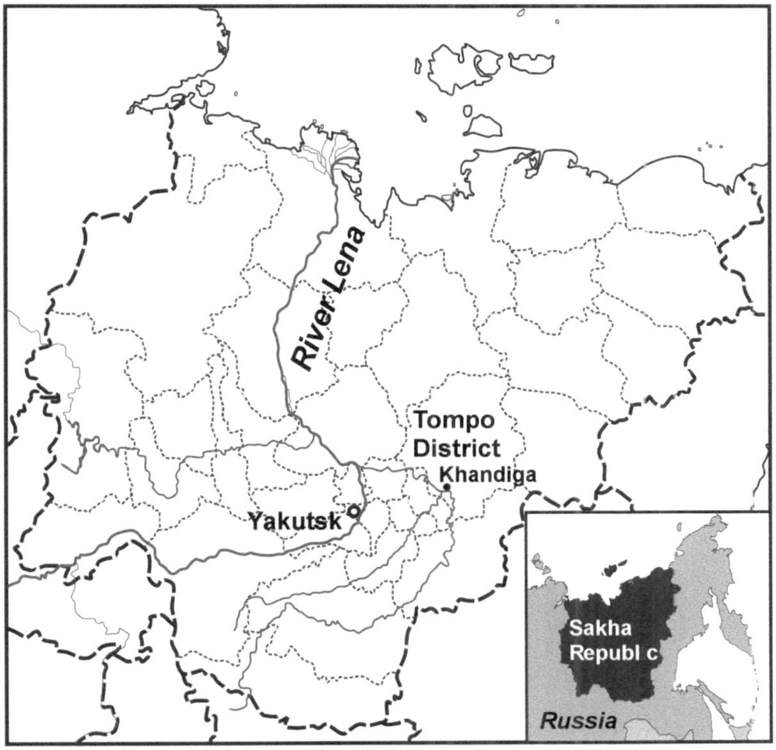

Fig. 10.1 Map of study site

and Klokov 2002). Among ~1,500,000 total domestic reindeer in Russia (Klokov 2012), 200,000 head are raised in Sakha (Ministry of Agriculture of Sakha Republic (Yakutia), Department of Northern Traditional Branch and Fishing 2010). Sakha is composed of a capital city (Yakutsk) and 34 districts; domestic reindeer are raised in 22 of these.

Tompo District is northeast of Yakutsk, and is one of the most successful reindeer herding districts in the republic. The distance from Yakutsk to the capital of the district, Khandiga, is about 400 km. The district area is 135,800 km^2, most of which is mountainous. Vegetation is classified as "mountain-taiga, sparse wood" (Plate 10.1). Mean air temperature is −39.7 °C in January and +16.2 °C in July. Mean precipitation is 16.1 mm in January and 100.9 mm in July (2007 data, Territorial Organ, Federal State Statistics Service in the Sakha Republic (Yakutia) and the Institute of Humanitarian Researches of the Academy of Sciences of the Sakha Republic (Yakutia) 2008). Population of the district is 15,300; 9200 live in the central town and 6100 in rural areas. Ethnic groups are as follows: Russians 48.3%, Sakha 34.5%, Even 5.4%, Evenki 0.5%, other 11.3% (Government of Sakha Republic (Yakutia) and the Institute of Humanitarian Researches of the Academy of Sciences of the Sakha Republic (Yakutia) 2007).

In Tompo District, about 170 workers raise 20,000 domestic reindeer in two organizations. Most of the reindeer are divided into 14 herds, and are managed by agricultural producer cooperative (trading station) "Tompo," originally state farm (*sovkhoz*) Tomponskii in the former Soviet regime. A herd of about 800 head is maintained by a clan community separated from the state farm in 1998.

Plate 10.1 Landscape of mountain taiga

10.3 Methods

First, to appraise the trend of reindeer herding in Tompo District, statistical data were used, i.e., fluctuations of domestic reindeer number in the entire Sakha Republic and individual reindeer herding districts from 1980 to 2010 (Ministry of Agriculture of Sakha Republic (Yakutia), Department of Northern Traditional Branch and Fishing 2010).

Second, climate change in the region was investigated using air temperature and precipitation data. These datasets were constructed as follows. In an area within the Tompo District (62.30–65.00°N, 135.00–141.00°E; Fig. 10.2), gridded data (at 0.5° latitude and longitude resolution) were built using spline interpolation of Baseline Meteorological Data in Siberia, provided by the Japan Agency for Marine-Earth Science and Technology. Then, monthly averages at each grid point were calculated for 1950–2008, and these values were averaged over the entire study area. Values larger than thrice the standard deviation were excluded as observational errors. Climate data were analyzed by seasons, i.e., spring (March, April, May), summer (June, July, August), fall (September, October, November), and winter (December, January, February).

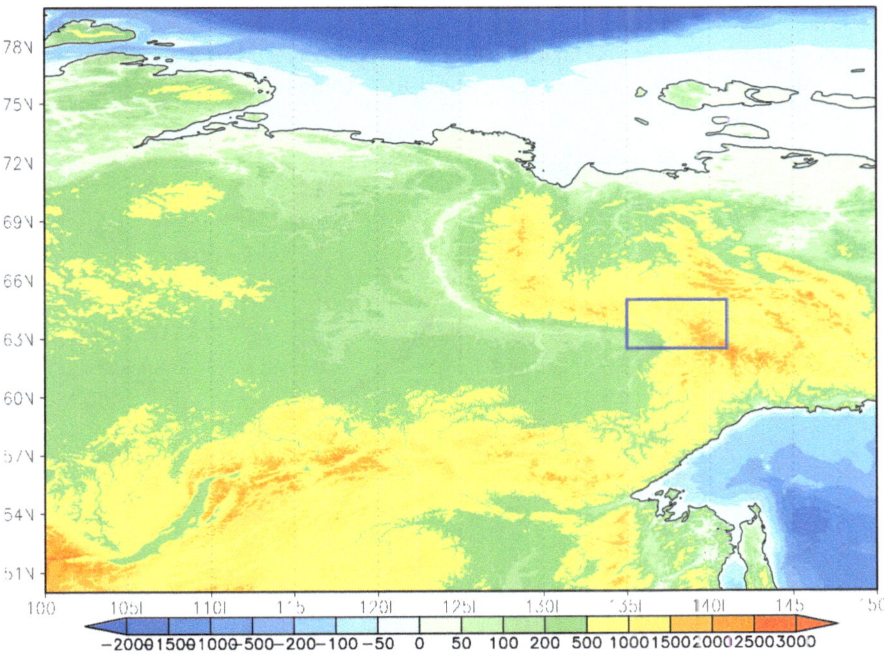

Fig. 10.2 Map of focus area. *Blue square* indicates area where climate data were obtained

Finally, field research was conducted at reindeer herding camps of the Even people during summer 2009 and autumn 2010. These camps belong to a clan community (*rodovaia obshchina*), a small organization formed by relatives of ethnic minority, in this case Even people. At the camp, some members of the clan community remain for several months in rotation and manage hundreds of reindeer. Each time I stayed with the people for about a week, conducting unstructured interviews with five herders (ages between the 20s and 50s) about climate, topography and other environmental change. I also participated and observed their subsistence activities when possible.

10.4 Results

10.4.1 Statistical Data

The statistical data show that the numbers of domestic reindeer in the entire Sakha Republic dropped sharply after the collapse of the Soviet regime, and that they have been recovering gradually in recent years (Fig. 10.3). Each district had various numerical changes of reindeer. For example, cases of the Olenek and Kobyai districts are shown along with that of the Tompo District (Fig. 10.3). Although the number of domestic reindeer in these three districts decreased similarly since around 1991 when the socialist regime lost power, their trends differed after that year. In

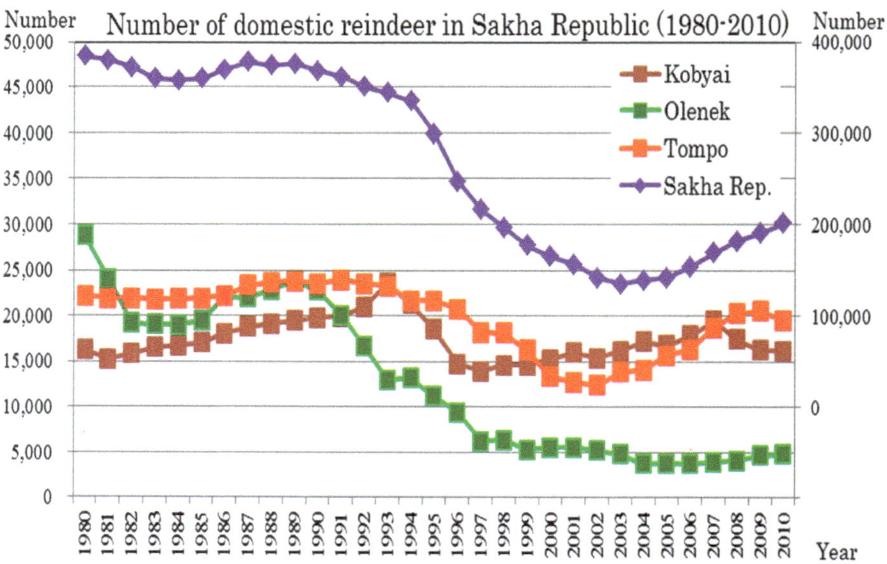

Fig. 10.3 Number of domestic reindeer in Sakha Republic and certain districts (1980–2010). *Left axis* indicates reindeer number in each district, and *right axis* that of the entire Sakha Republic

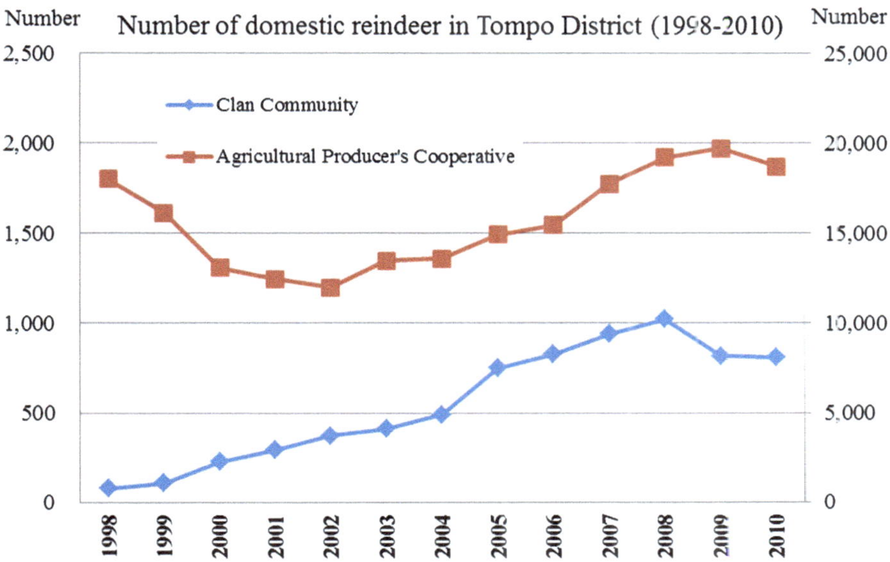

Fig. 10.4 Number of domestic reindeer in Tompo District (1998–2010). *Left axis* indicates reindeer number of clan community, and *right axis* that of agriculture producer's cooperative

Tompo and Kobyai districts, the tendencies were similar to the entire Sakha Republic, and reindeer numbers have been recovering recently. However, Olenek District differed from other districts and its trend became steady at a small number. Figure 10.4 shows numerical changes in domestic reindeer of the two organizations in Tompo District from 1998 to 2010. The clan community was established in 1998, after which the number of reindeer gradually increased (Department of Peoples Affairs and Relationships with Federation of Sakha Republic (Yakutia) 2008), as was the case for the agriculture producer's cooperative. These data indicate that reindeer herding in the district has been successful in recent years.

10.4.2 Climatic Data: Air Temperature and Precipitation

The meteorological data indicate that air temperature and precipitation in the area varied interannually. Annual mean temperature is gradually rising. In each season, mean temperatures fluctuated between −14.9 and −6.6 °C in spring, 11.5 and 15.7 °C in summer, −17.8 and −9.5 °C in autumn, −46.4 and −36.1 °C in winter, and tended to gradually increase (Fig. 10.5). However, the interior of the Sakha Republic is well known for extremely low temperatures in winter, which remain below −35 °C during the winter months.

Annual precipitation is generally stable on a long-term basis (~60 years), but seasonal variation of precipitation was substantial, especially in autumn and sum-

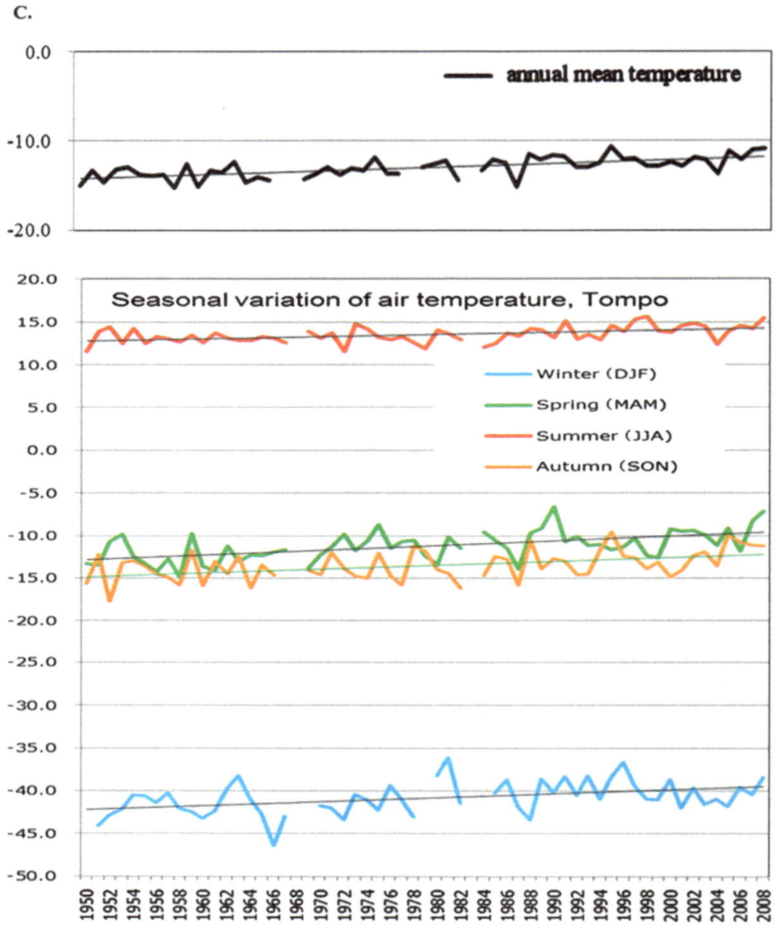

Fig. 10.5 Seasonal variation of air temperature in Tompo District

mer (Fig. 10.6). The temperature tended to increase markedly in summer but was nearly constant in winter. This means that snow, which can obstruct forage of reindeer in winter, is not increasing.

10.4.3 Field Research

According to my field research, herd management is done daily as follows. In the morning, reindeer that have roamed freely at night are collected at the campsite by a herder (Plate 10.2). In the daytime, reindeer are kept at the campsite and surrounding area. A herder prevents them from leaving the area. At night, the reindeer are

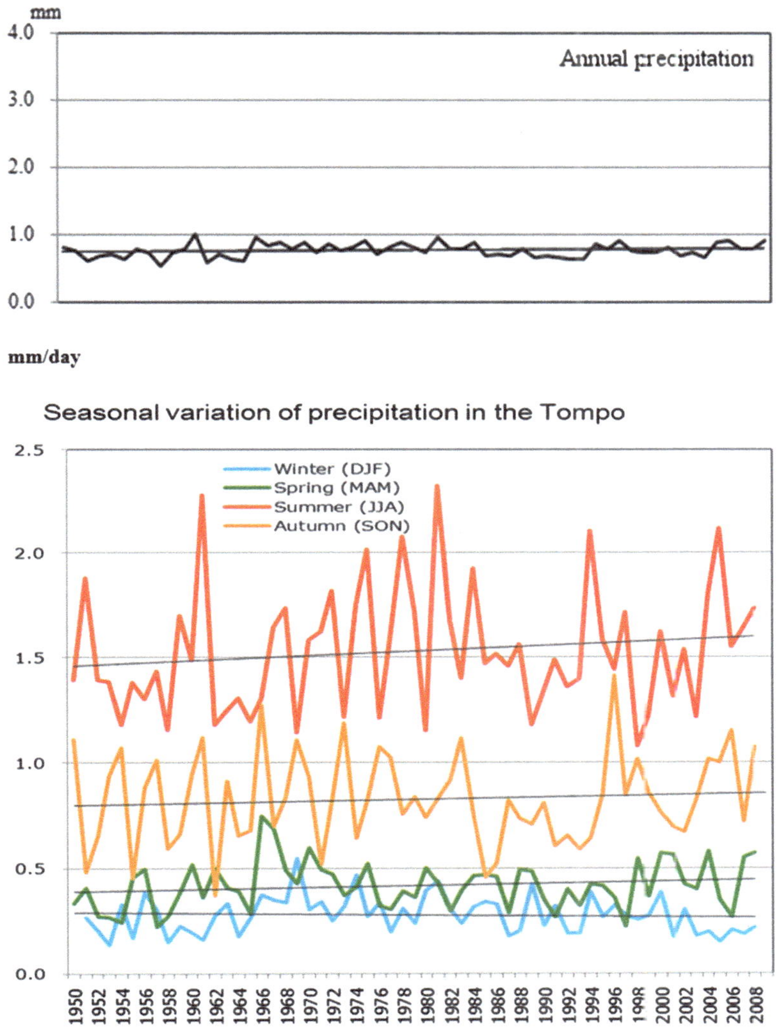

Fig. 10.6 Seasonal variation of precipitation in Tompo District

allowed to roam freely. Because a similar pattern was reported in the Eveno-Bytantaysky District of the northern Sakha Republic (Takakura 2004), this appears to be a general form of reindeer herding in the region.

Although this form of herding seems to be simple and easy, herders used certain characteristic techniques to manipulate reindeer efficiently. Monitoring reindeer by bells is one of these techniques. About 5–10% individuals in the herd wear collars with a bell (Plate 10.3). People at the camp pay attention to the tinkling bells, whereby they can track approximate locations and movements of the herd. Another technique is attracting reindeer by salt. At the campsite, salt is placed on a type of

Plate 10.2 Reindeer gathered together at campsite by a herder

Plate 10.3 Some reindeer wear collars with a bell

stand near a tent so that the reindeer can ingest it *ad libitum*. Sometimes people give reindeer salt directly by hand. The salt lures the reindeer and helps keep them near the campsite. There are other methods to manage reindeer. When herders want to catch reindeer, they use lassos. Sometimes a dog is used to drive and collect reindeer, and this appears to aid the herder's task.

Herders and other members of the clan community use domestic reindeer as food. Reindeer meat is partly consumed in their daily meals and are partly offered for sale. Reindeer milk is an occasional delicacy for the people. They use reindeer fur and leather for materials to make clothing and goods for daily use. Riding and pulling sledges are also important roles for reindeer (Plate 10.4).

Plate 10.4 Riding reindeer

Generally, herders stay at the campsite for several months. Besides herding reindeer, they hunt for wild animals such as Siberian bighorn, wild reindeer, elk, sable, grouse and others. They also fish and gather wild berries and mushrooms. Because meat is the principal item in their diet, hunting is especially important for subsistence and, if they cannot catch wild animals, they must slaughter their domestic reindeer to obtain meat. Thus, they can conserve their livestock by hunting wild animals.

Ventsel (2006) insisted that for indigenous people of the Anabar District in the northern Sakha Republic, hunting and reindeer herding can be seen as complementary strategies, and that the people have lived in the hunter–herder continuum. From his field research of Even and Nenets reindeer herders, Takakura (2010) argued that when any social organization faces certain ecological constraints and socioeconomic-political settings, it adaptively manages to invent a particular combination of subsistence choices, i.e., a "subsistence continuum". Thus, the combination of reindeer herding and hunting in the present research appears not to be a special case but a general state, at least for taiga reindeer husbandry.

Herders of the Even clan community in Tompo District managed reindeer herds largely in the traditional manner. They formed a few reindeer herding parties, herded in remote natural pasture year round, and managed herds using traditional techniques. In natural pasture, reindeer chiefly forage plants and lichen. Consequently, reindeer and their herders must be sensitive to environmental change. And naturally, they have some information about global warming from the mass media.

Herders of the clan community sensed fluctuation of climatic variables such as air temperature, precipitation, and snow depth. However, their remarks about climatic, environmental and topographical changes were rare and limited. A most frequently heard remark was that there had hardly been any influence of global warming in the area.

In August 2009, an interview was held with reindeer herders Sergei (male, 39 years old; all names are pseudonyms) and Peter (male, 25 years old). Peter is an older brother of Sergei's wife, and they lived in a tent and herded reindeer together in cooperation with another family of the clan community. They said that they did not feel any influence of global warming in the area. Sergei said that if global warming did affect the area and increased air temperature, he would be concerned that reindeer would not forage sufficiently and lose weight. He felt that their important game, such as Siberian bighorn and Arctic graylings, have been decreasing, but did not connect this with global warming. As mentioned above, if there is ample game, the people do not have to slaughter domestic reindeer for food. The population of wild game directly affects the consumption volume of domestic reindeer by locals. In September 2010, another interview was conducted with other herders in the clan community: Boris (male, 50 years old), his wife Eva (46 years old), and their son Stephen (25 years old). Boris and Eva were representative members of the clan community from its foundation, and Boris is an older brother of Sergei. They set up a tent near a pasture and lived with a younger brother of Stephen (13 years old; he was thought too young for the interview and was therefore excluded), managing a part of the reindeer herd of the clan community (about 70 head). These people stated that they were unaware of any climatic and environmental changes, such as rainfall, snow depth, or flood frequency. However, Stephen felt that landslides around their pastures had increased in recent years (Plate 10.5). Comments from the interviewees on environmental change are summarized in Table 10.1.

Plate 10.5 Landslide, the type of which a herder remarked has recently increased

Table 10.1 Informants of the field research

Fictitious name	Age	Sex	Relationship	Comments on environmental change
Sergei	39	M	Boris's younger brother	Decrease of game
Peter	25	M	Sergei's brother by marriage	No change
Boris	50	M	Sergei's older brother	No change
Eva	46	F	Boris's wife	No change
Stephen	25	M	Son of Boris and Eva	Increase of landslides

10.5 Discussion

The statistical data show that the trend of domestic reindeer number in Tompo District was similar to the entire Sakha Republic and Russian Federation. The situation varied by region in Russia. For example, reindeer herding grew in some places of the western Siberia tundra zone, even after the Soviet regime collapsed (Yoshida 2001; Forbes et al. 2009). Generally, however, numbers of domestic reindeer declined sharply after that collapse, and have gradually recovered in recent years. In a village of Olenek District in the northwestern Sakha Republic, reindeer herders shifted their subsistence activity to reindeer hunting (Takakura 2012). Domesticated reindeer of Chukchi decreased considerably, and reindeer herding in the taiga regions declined throughout Russia after the collapse (Syroechkovski 2000).

The major reduction of reindeer numbers after the collapse of the Soviet regime is considered to be associated with a withdrawal of government subsidies (Baskin 2005; Klokov 2012). Although the Soviet regime provided considerable support to reindeer herders such as free veterinary assistance, helicopter rentals at low prices, and social welfare, these were withdrawn as the regime collapsed. From the end of the 1990s, with stabilization of the Russian economy, the number of reindeer increased gradually. This was because of development of the market economy, initiation of support from regional government, and improvement of legal frameworks in support of independent reindeer households. Thus, Klokov (2012) suggested that changes in the total number of domestic reindeer in Russia can be most closely connected with changes of political context.

Klokov also suggested the possibility of some connection between reindeer number and climatic drivers. However, climatic influences on domestic reindeer numbers have rarely been investigated. In this chapter, I have focused on relationships between climatic indicators and those numbers. The resulting trends indicate no direct and/or indirect influence of environmental change on reindeer number, not only for all Russia but also for the Sakha Republic and Tompo District, at least through the present. Although there might have been some negative or positive influence of environmental change, social and political changes appear to be much more critical to reindeer herding.

The variation of climatic data shows that air temperature and precipitation are gradually increasing, but interannual variations are large (Figs. 10.5 and 10.6). Some herders recognize little environmental change. However, it appears that they take for granted that climatic variables fluctuate, and believe that this fluctuation has been within normal limits. The herders generally do not report abnormal environmental changes. Oskal et al. (2009) reported on reindeer herder recognition of their challenges, including climate change, at eight locations in Norway, Finland and Russia. Similar to the present study, climate change issues were not seen as a major challenge by herders at Topolinoe village (Tompo District) in their report. That is, though herders noticed that climate was changing, they appeared unconcerned about it.

These results indicate that although local weather has been changing, it was not a serious matter to reindeer herders in the Tompo District during 2009 and 2010. There appears to be several reasons for this disregard of climate change. The first is that change of air temperature and precipitation is gradual, and there is a much wider annual fluctuation. Thus, even herders who work in the field for a long period may not clearly perceive the change. Second, the fauna, flora, and topography in the Tompo District appears not to have changed considerably, and reindeer herding has not been deleteriously affected. Precipitation is nearly constant during the winter months. Because of this, in conjunction with the low temperatures, neither snow depth nor ice crust (which can disturb reindeer forage) have increased in the Tompo District. Because winter temperature in the region is extremely low (about −40 °C), a slight increase of this temperature may not critically influence the natural environment. According to Oskal et al. (2009), the Tompo District is much colder in winter than the other reindeer herding areas in their report; mean winter temperatures (1961–1990) were −16.0 °C at Kautokeino in Norway, −22.9 °C at Salekhald in the Yamal-Nenets Autonomous District of Russia, and −20.7 °C at Anadyr in the Russian Chukotka Autonomous District. Third, social and/or economic factors may be more important to reindeer herding, at least in Russia (Forbes and Stammler 2009; Klokov 2012). There was a serious decline of reindeer number at the time of Soviet regime collapse. After the market economy became established in Russia, because reindeer herding was supported by subsidies from the Sakha Republic and Russian Federation, the number of domestic reindeer gradually increased in both the republic and Tompo District. This relationship is supported by the fact that reindeer numbers increased with investment in the republic over 2002–2009 (Fig. 10.7).

In summary, despite gradual climatic and environmental change, because air temperature is sufficiently cold and precipitation is stable (so that snow depth and ice crust frequency do not increase), reindeer herding appears resilient to such change. Moreover, because reindeer herding has been effective with sufficient subsidies, herders appear unconcerned about the influences of that change.

Global warming may generally harm reindeer herding, but its effect is not uniform across locations. Further, this climate change may impair herding through the formation of ice crust on snow in northern Europe and western Siberia, and by reindeer-drinking-water shortages in Chukotka. The impairment may occur indirectly through acceleration of human development activities, devastating pastures (Oskal 2008; Oskal et al. 2009). However, the effects of global warming are not necessarily harmful, and can be somewhat favorable (Baskin 2005; Klokov 2012). For example, Klokov (2012) indicated that during 1960–1990, there was slow growth in reindeer numbers in northeastern Russia because of better summer pasture conditions caused by climatic change.

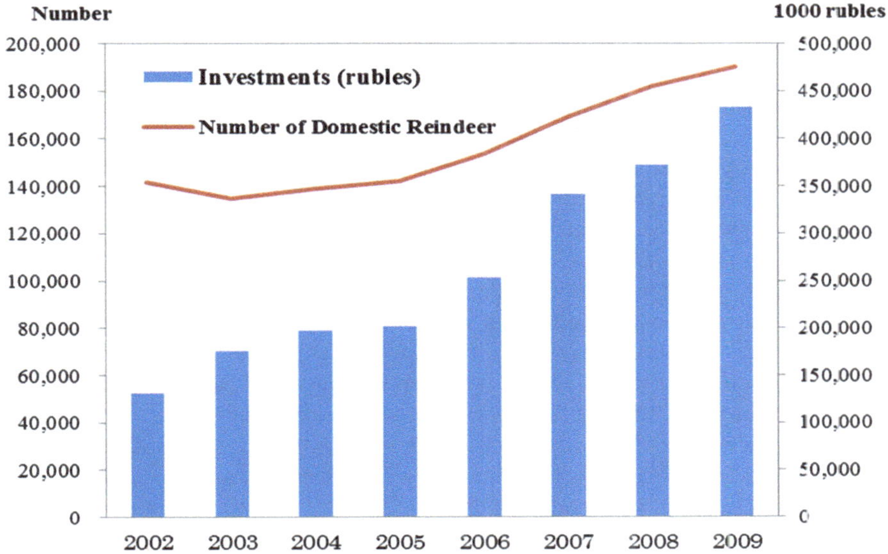

Fig. 10.7 Relationships between amount of investments and number of domestic reindeer in Sakha Republic (Source: Ministry of Agriculture of Sakha Republic (Yakutia), Department of Northern Traditional Branch and Fishing 2010)

In the present research, climate change appears not to have harmed reindeer herding in the Tompo District of the Sakha Republic. However, Yoshida (2012b) reported that in the Kobyai District (adjoining the Tompo District), climate change was perceived to have a negative influence on reindeer herding; repeated thawing–freezing of snow which make ice crust affected the mortality of reindeer calves. These phenomena were not found in the present study. However, requirements for the occurrence of ice crust are unclear, and air temperature in the Tompo District during spring (calving season) does not differ much from that in the Kobyai District.

Thus, to verify the relationships between climate, environmental change, and reindeer herding, we require more detailed information regarding the effects of climate change on the natural environment in the study area, together with comparative studies of areas where there has been strong climate change and environmental damage.

10.6 Conclusions

Climatic and environmental changes induced by global warming may directly and indirectly influence populations of wild and domestic reindeer throughout the north. In the present study, climatic changes were evidenced. However, a negative impact of such changes was not detected either in the variation of domestic reindeer numbers or in the awareness of herders.

The reasons for these findings are as follows. First, compared with the much wider annual fluctuation of climatic variables, the increase of air temperature and precipitation are so slight that herders cannot perceive them. Second, it appears that neither the natural environment nor reindeer herding in the Tompo District have been significantly damaged. The observed climate change in the district was a gradual increase of air temperature. However, because this temperature has remained extremely low and precipitation is relatively stable in winter, the amount and density of snow and the occurrence of ice crust that obstruct reindeer foraging may not be increasing. Third, the Russian Federation and Sakha Republic have subsidized reindeer herding, and the number of domestic reindeer has increased in recent years. Therefore, current reindeer herding in Tompo appears to be managed satisfactorily, and herders are not conscious or anxious about subtle changes.

Although climate change induced by global warming may harm reindeer herding in general, this effect may not be uniform across locations. The results of the present research indicate that climate change appears not to have harmed reindeer herding, at least in the Tompo District. However, there have been contradictory results from studies in areas near the district. Subtle differences of climatic or geographic features may modify each situation. To explain these differences and verify the relationships between environmental variables and reindeer herding, we require more detailed studies of climatic and environmental changes in the region, together with comparative studies of neighboring areas.

Acknowledgments We gratefully acknowledge V. D. Ingat'eva, S. I. Boyakova, and the Institute of the Humanities and Indigenous Peoples of the North of the Siberian Branch of the Russian Academy of Sciences, for their support and assistance throughout the study. We appreciate Drs. K. Yamamoto and K. Oshima for preparing datasets of climate variables. We also express gratitude to M. P. Pogodaeva, M. P. Pogodaev, and all herders in the Pogodaev clan community for acceptance and cooperation with the field research.

References

ACIA (Arctic Climate Impact Assessment) (2004) Impacts of a warming arctic: arctic climate impact assessment. Cambridge University Press, Cambridge

Baskin LM (2005) Number of wild and domestic reindeer in Russia in the late 20th century. Rangifer 25(1):51–57

Department po delam narodov i federativnym otnoschenijam Respubliki Sakha (Yakutia) (2008) Mir olenevodov. Department po delam narodov i federativnym otnoschenijam Respubliki Sakha (Yakutia). (In Russian)

Forbes BC, Stammler F (2009) Arctic climate change discourse: the contrasting politics of research agendas in the West and Russia. Polar Res 28:28–42

Forbes BC, Stammler F, Kumpula T, Meschtyb N, Pajunen A, Kaarlejarvi E (2009) High resilience in the Yamal-Nenets social-ecological system, West Siberian Arctic, Russia. Proc Natl Acad Sci U S A 106(52):22041–22208

Gordon B (2003) Rangifer and man: an ancient relationship. Rangifer, Special Issue 14:15–28

Grayson DK, Delpech F (2005) Pleistocene Reindeer and global warming. Conserv Biol 19(2):557–562

IPCC (Intergovernmental Panel on Climate Change) (2007) Climate change 2007: synthesis report. IPCC

Jernsletten J-L L, Klokov K (2002) Sustainable reindeer husbandry. Arctic Council 2000–2002: Centre for Saami Studies, University Tromso

Klein DR (1991) Limiting factors in caribou population ecology. Rangifer, Special Issue No.7:30–35

Klokov KB (2004) Russia. In: Ulvevadet B, Klokov K (eds) Family-based Reindeer herding and hunting economies, and the status and management of wild Reindeer/Caribou populations. Centre for Saami Studies, University of Tromso, pp 55–94

Klokov KB (2012) Changes in reindeer population numbers in Russia: an effect of the political context or of climate? Rangifer 32(1):19–33

Ministerstvo sel'skogo khozjaistva Respubliki Sakha (Yakutia) Department traditsionnykh otraslei Severa i rybokhozjaistvennoi dejatel'nosti (2010) Domaschnee olenevodstvo Respubliki Sakha (Yakutia). Ministerstvo sel'skogo khozjaistva Respubliki Sakha (Yakutia) Department traditsionnykh otraslei Severa i rybokhozjaistvennoi dejatel'nosti, Yakutsk. (In Russian)

Oskal A (2008) Old livelihoods in new weather: arctic indigenous reindeer herders face the challenges of climate change. Dev Outreach 10(1):22–25. World Bank Institute

Oskal A, Turi JM, Mathiesen S, Burgess P (eds) (2009) Ealat Reindeer Herders' voice: Reindeer herding, traditional knowledge and adaptation to climate change and loss of grazing land. International Centre for Reindeer Husbandry

Pravitel'stvo Respubliki Sakha (Yakutia) i Institut gumanitarnykh issledovanii AN Respubliki Sakha (Yakutia) (2007) Yakutia: Istoriko-kul'turnyi atlas. Izdatel'stvo Feoriia, Moscow. (In Russian)

Rees WG, Williams M, Vitebsky P (2003) Mapping land cover change in a reindeer herding area of the Russian Arctic using Landsat TM and ETM+ imagery and indigenous knowledge. Remote Sens Environ 85:441–452

Rees WG, Stammler FM, Danks FS, Vitebsky P (2008) Vulnerability of European reindeer husbandry to global change. Clim Chang 87:199–217

Sasaki S (1984) Reindeer nomadism in Siberia: a study of the case about Nenets in West Siberia and its economical meaning. Quart J Anthropol 15(3):114–180. (In Japanese)

Syroechkovski EE (2000) Wild and semi-domesticated reindeer in Russia: status, population dynamics and trends under the present social and economic conditions. Rangifer 20:113–126

Takakura H (2004) Gathering and releasing animals: reindeer herd control activities of the indigenous peoples of the Verkhoyansky region, Siberia. Bull Nat Museum Ethnol 29(1):43–70

Takakura H (2010) Arctic pastoralism in a subsistence continuum: a strategy for differentiating familiarity with animals. In: Stammler F, Takakura H (eds) Good to eat, good to live with: nomads and animals in Northern Eurasia and Africa. CNEAS, Tohoku Unieversity, Sendai, pp 21–42

Takakura H (2012) The shift from herding to hunting among the Siberian Evenki: indigenous knowledge and subsistence change in Norhwestern Yakutia. Asian Ethnol 71(1):31–47

Territorial'nyi organ federal'noi slujby gosudarstvennoy statistiki po Respublike Sakha (Yakutia) (2008) Ekonomika rayonov Respubliki Sakha (Yakutia) 1900, 2000–2007 gg.: Statisticheskiy sbornik. Territorial'nyi organ federal'noi slujby gosudarstvennoy statistiki po Respublike Sakha (Yakutia), Yakutsk. (In Russian)

Vainshtein S (1980) Nomads of South Siberia: the pastoral economies of Tuva. Translated by Michael Colenso. Cambridge University Press, Cambridge

Ventsel A (2006) Hunter-herder continuum in Anabarski district, NW Sakha, Siberia, Russian Federation. Nomadic People 10(2):68–86

Weladji RB, Forbes BC (2002) Disturbance effects of human activities on *Rangifer tarandus* habitat: implications for life history and population dynamics. Polar Geogr 26(3):171–186

Weladji RB, Holand O (2006) Influences of large-scale climatic variability on reindeer population dynamics: implications for reindeer husbandry in Norway. Clim Res 32:119–127

Weladji RB, Klein DR, Holand O, Mysterud A (2002) Comparative response of Rangifer tarandus and other northern ungulates to climatic variability. Rangifer 22(1):33–50

Wolfe SA, Griffith B, Wolfe CAG (2000) Response of reindeer and caribou to human activities. Polar Res 19(1):63–73

Yoshida A (2001) Some characteristics of the tundra Nenets reindeer herders of western Siberia and their social adaptation. In: Anderson DG, Ikeya K (eds) Parks, property, and power: managing hunting practice and identity within state policy regimes, Senri Ethnological Studies no. 59. National Museum of Ethnology, Osaka, pp 67–80

Yoshida A (2012a) Reindeer herding/husbandry and problems of development and environment in Siberia. In: Takakura H (ed) Living in severe cold Siberia: reindeer, ice and Indigenous peoples. Shinsensha, Tokyo, pp 137–156. (In Japanese)

Yoshida A (2012b) Reindeer herding and environmental change in Kobyai and Olenek districts, Sakha Republic. In: Proceedings of 1st international conference on "Global warming and the human-nature dimension in Siberia: social adaptation to the changes of the terrestrial ecosystem, with an emphasis on water environments". pp 92–96

Chapter 11
History of Transport Infrastructure Development in the Sakha Republic (Yakutia)

Sardana I. Boyakova

Abstract The goal of the article consists in discussing the features of the formation of the transport system in the Sakha Republic (Yakutia), the modern state of the main sectors of Transport and Communications of the Republic, new risks for their safe operation, arising from global climatic changes. The article considers in detail history of the formation and development of the transport system of Yakutia. More specifically, the following types of transport are discussed: river and sea navigation, aviation, railway transportation, automobile transportation and delivery of natural resources by pipelines. Special attention is paid to such peculiar seasonal logistics systems as winter roads, also referred as ice roads (zimnik's). The author analyzes particular features of the functioning of these roads in the conditions of the North, construction and state of ice crossings and their roles in regional economy. The article emphasizes the need to develop measures for cushioning the negative impact of climatic change on the region's transport infrastructure.

Keywords Transport • History • Climate change • Livelihood • Winter roads (zimnik) • Security • Waterways • Risks • Damage assessment

11.1 Introduction

Worldwide climate changes have been recognized not only as an environmental problem but also as a problem of the economy and human security. Changes in the environment affect the functioning of those areas, which are sustained and limited by the components of ecosystems, i.e., agriculture, forestry, water management, recreation and leisure, energy economics, construction, transport, and mining (Khomyakov et al. 2005). In these endeavors, worldwide climate change can affect water ecosystems both positively and negatively. However, extreme climate

S.I. Boyakova (✉)
The Institute for Humanities Research and Indigenous Studies of the North of the Siberian Branch of the Russian Academy of Sciences, 1 Petrovskaya str., Yakutsk 677027, Russia
e-mail: boyakova@mail.ru

T. Hiyama, H. Takakura (eds.), *Global Warming and Human - Nature Dimension in Northern Eurasia*, Global Environmental Studies,
DOI 10.1007/978-981-10-4648-3_11

events and other abrupt deviations that reduce the time for adaptation are likely to result in more severe impacts on all societal systems, including transport.

For Siberia and especially its northeast, logistics communications have always been developing, and any change in the conditions under which these communications are constructed and exploited can complicate the functioning of nearly all branches of the economy. Transport is a relevant component of the contemporary economy of Yakutia, because it ensures the transfer of essential cargo, goods, and people. It is significant in safeguarding the quality of life and comfortable living standards in the north, and provides continuous supplies to such strategic sectors of the economy as mining. The transport network not only satisfies human and economic needs but forms a material base for socioeconomic development of the region.

In this chapter, I focus on how Yakutia's transport infrastructure has evolved, its modern state, possible consequences of climate change for the infrastructure, and risks in connection with these consequences posed to the regional livelihoods of the people. The work is based on archive documents from Moscow, Saint Petersburg, Vladivostok, and Yakutsk, as well as literary sources, statistical data, and field materials gathered in Zhiganskii, Megino-Khangalaskii and Bulunskii uluses (districts) of the Sakha Republic (Yakutia) in 2008–2011.

11.2 Natural Climatic Conditions and Topography

The modern territory of Yakutia occupies the northeastern part of the Eurasian continent and has 3.1032 million km^2 of total area, including adjacent islands of the Arctic Ocean. The area is over 2000 km from north to south and over 2500 km from west to east.

Yakutia is a mostly mountainous region; almost two thirds of its area is occupied by mountains and plateaus. Several mountain ranges, mainly meridional in direction, traverse the entire northeastern part of the region. There is a general downward slope from south to north, in the direction of the Laptev and East Siberian seas. Four landscape zones can be distinguished within Yakutia; the majority of this vast space (almost 80%) is covered by taiga forest, and the remainder by tundra, forest tundra, and arctic deserts (Table 11.1).

In Yakutia, around 650,000 watercourses over 10 km long have been documented. These include 617 medium rivers (100–500 km long) with overall length 115,000 km, and 53 large rivers (> 500 km long) extending 48,700 km (Drachev et al. 2005). Large rivers in the republic are the Lena (4400 km), Kolyma (2129 km), Indigirka (1726 km), Vitim (1978 km), Olenyok (2292 km), Olyokma (1436 km), Vilyuy (2650 km), and Aldan (2273 km), which traverse the entire region and empty into the Arctic Ocean. Basins of the lower and middle reaches of the first three rivers are dominated by extensive lowlands, namely, the North Siberian, Yana-Indigirka, and Kolyma. These are characterized by an abundance of lakes that are nearly continuous; these are linked with each other and not infrequently with the nearest large rivers, small grassy creeks, and brooks. Many lakes have areas of hundreds of square kilometers, for example, Nidjili, Labynkyr, Ozhogino, Chukochie, and Nerpichie.

Table 11.1 Effects of Relief, Nature, and Climate on Transport in Sakha Republic (Yakutia) (Drachev et al. 2005)

Index	Plain	Lowland	Highland
Altitudinal zone, %	8	60	32
Average altitude, m	65	510	1200
Days per year with temperatures equal or below, n			
0°	285	179	256
−15°	179	150	164
−25°	95	110	117
−35°	29	62	69
Annual precipitation, mm	140–150	175–300	190–600
Snow cover depth, cm	20–120	20–75	20–75
Days with stable snow cover, n	230	220	225
Snowstorm duration per year, hours	до 690	до 220	до 1100
Snowdrift per 1 m of road, m^3	до 1000	до 200	до 800
Ratio of average technical speed of motor transport	1.00	0.87	0.60
Ratio of 1 km of road construction	2.2	1.0	1.4
Water barriers per 100 km of ordinary and winter roads, n	13	12	18
Rivers freezing-over duration, days	240–245	210–235	200–240
Duration of ice thickness, days			
40 cm	190–210	130–205	145–210
80 cm	40–190	20–170	60–190

The lowlands are frequently covered for tens of kilometers by swamps (*badarans*), which are impassable in summer. The installation of transport communications is considerably complicated by permafrost and a sharply continental climate, characterized by long, cold winters and short, warm summers. There is snow cover from 6.5 to 9 months per year. Yakutia bounds a sea area of as much as 800,000 km^2, which embraces the northern sea route along the shore. In the eastern Arctic, sea navigation begins about 1–2 months later than river navigation, but both close almost contemporaneously.

Natural-climatic conditions and the landscape of the region determine its dispersed character and the wide geographic resettlement of the peoples. The low population density and nomadic lifestyle of some indigenous peoples made the construction of roads unnecessary as all movement and transportation of loads were done using pack animals.

11.3 Construction of the Transportation System

The annexation of Yakutia to Russia necessitated the creation of the communications system in the region, primarily for closer ties with the (Russian) center. Already by the seventeenth century, the Cossacks had built water and land routes between *ostrogs* (fortresses) and wintering places. In 1743, the Irkutsk–Yakutsk highway was founded, which became a continuation of the main Siberian official highway, with the point of departure at the imperial capital of Saint Petersburg. Travel on this highway along the Lena River was possible year round, thanks to stations built at regular intervals. The winter road passed over river ice and on the shore; in the summer, the waterway was used. For transportation of loads and communication with the Pacific coast and Russian colonies in North America, the Okhotsk highway was laid in the second half of the eighteenth century. In the mid-nineteenth century, the Ayan highway was built. Movement along these routes was also year round, except in the coldest months (December–February) when temperatures dropped to −50 to −60 °C.

In the same period, regular navigation in the region began. The first steamship on the Lena River came into use in 1856, and by 1917 there were already 38 ships. In 1878, the steamboat *Lena* arrived at the estuary of that river from Sweden, carrying the expedition of A.E. Nordenskiöld who was making a historic voyage along the northern sea route. Later, the steamboat was used for communication between the city of Yakutsk and lower reaches of the Lena. At the end of the nineteenth and beginning of the twentieth centuries, the tributaries of the Lena, the Vitim, Vilyuy, Olyokma, and Maya, had been mastered. From 1911, sailing began along the eastern part of the northern sea route. By order of the Russian government, the Dobroflot company made the first voyage from Vladivostok to the Kolyma estuary. Subsequently, these voyages became yearly. The intention was to extend them to the Lena estuary but the plans were interrupted by the First World War and the Russian revolution. During the civil war, Kolyma voyages were suspended and the purveyance of goods to the northern districts of Yakutia passed to Americans, who had also opened a seaway to Kolyma in 1911.

Thus, the formation of Yakutia's transportation system was based on the natural idiosyncrasies of the region, with waterways at its base. Land roads had a mixed character; in winter these partially crossed river ice, and in summer some portions involved the river. The construction of transport communications was determined by economic and political priorities. Thus, the development of navigation on the Lena River was directly connected with the initial development of gold mining in the second half of the nineteenth century (Boyakova 2001).

Rapid development of air transport began in the Soviet era. From 1925, when the first airplane arrived at Yakutia through the early 1990s, the entire republic was covered by a network of airfields. Small aircraft aviation underwent rapid development and An-2's (small, single-engine biplanes) flew several times weekly, even to small villages. For communication with remote camps or other remote areas, helicopter facilities were active, based in Magan, Batagay, Nyurba, and Tiksi. Therefore, almost no roads were built, and major efforts were directed not at internal road construction but at building roads externally to connect Yakutia to neighboring regions. For winter delivery of goods, a route from the village of Big Never to Yakutsk was paved, which connected the republic with the Siberian railroad. Simultaneously, the delivery of goods to Yakutia via the northern sea route began in 1933. The development of this main artery promoted navigation on the Yana,

Indigirka, Kolyma, Olenyok, and Anabar Rivers, as well as the control of transport routes through cabotage. In the 1970s in connection with the development of coal deposits in southern Yakutia, the region was reached by railroad. A branch was built from the Baikal-Amur main line to Chulman and Berkakit. Railroad construction continued in the post-Soviet period and, in 2011, the railroad reached its final destination in the village of Nizhny Bestiakh.

11.4 Current State of Transport Infrastructure

Currently, all types of transport exist in Yakutia, including railway, air, motor, water (sea and river), and pipeline traffic (Fig. 11.1).

As described above, railway transport is the most recent in Yakutia. Its development in the republic will generate opportunities for enhancing regional economic growth, successful exploitation of mineral deposits in southern Yakutia, expansion of the year-round transportation access zone over a considerable part of the republic (with the major part of its population), and cost reduction for goods delivery.

Fig. 11.1 Transport infrastructure of the Sakha Republic (Yakutia) at the beginning of the twenty-first century

The planned construction of a transport passage (bridge or tunnel) across the Lena River will connect the capital of the republic, Yakutsk, with the all-Russian railway network. In southeastern Yakutia, construction of the terminal Ulak-El'ga rail line will connect the Baykal-Amur Railroad with the El'ga coal deposit.

Yakutia is still among the Russian regions with relatively developed air traffic. By the early 2010s, there were 23 functional airports in the republic. Air transport carries the most passengers in the republic, ~65% of all passenger traffic. Besides long-haul passenger traffic, there is short-distance transportation between settlements and districts inside the republic. Air transport is used for medical and other services, aiding people in remote localities, extinguishing forest fires, destroying ice blocks, and conducting aerial surveying and reconnaissance.

Yakutia has three federal airports (Yakutsk, Tiksi, and Chokurdakh), 35 local airfields (including nonoperational ones), and 195 landing fields (airstrips without facilities for air traffic control). Nine airports are equipped with paved runways, Yakutsk, Mirny, Chokurdakh, Cherskii, Viliuysk, Chul'man, Aldan, and Poliarny. There are immediate plans to close some airports and renovate 16, constructing new facilities at eight of these. The major airports have air communication with central Russia. Yakutsk airport is directly connected with Moscow, Saint Petersburg, Novosibirsk, Krasnoyarsk, Irkutsk, Khabarovsk, Magadan, and other cities and towns. It also has international flights to 13 countries.

Water transportation is now the major mode in Yakutia, and is the primary means of delivering supplies to the population of the region. The primary goods deliveries are conducted during the short period when northern navigation is possible (4.5 months, early June through mid-October), both outside and inside the republic. Exploited waterways in Yakutia extend 21,800 km, and operating waterways 13,600 km. Yakutia has an extensive composite water transport network: A segment of the northern sea route is the major waterway of the Lena River, which flows across the entire republic with its tributaries the Aldan and Viliuy, and the navigable Anabar, Olenek, Yana, Indigirka, and Kolyma rivers in northern Yakutia. Sea transportation operates in 12 Arctic districts, from Khatanga Bay to the port of Pevek and the Chukot regions.

The water transportation system of the republic comprises five river ports (Lensk, Olekmink, Yakutsk, Nizhneyansk, and Belaya Gora) and two Arctic seaports, Tiksi and Zeleny Mys. Republic maritime transport features a combined sea–river fleet, with oil and bulk carriers and other vessels. Thus, the Lena Associated River Shipping Company can transport cargo along the Lena River, its tributaries, and other rivers of Yakutia, as well as conduct sea trade on the coast.

The major portion of goods is currently transported to the republic through the combined railway/water route of the Trans-Siberian Railroad, i.e., from the Baykal-Amur Railway to the port Osetrovoto to the Lena River. The port of Osetrovo (Ust'-Kut town, Irkutsk Oblast) does 80% of the cargo handling in the republic. This stable, decades-old transportation pattern may gradually change with completion of the Amur-Yakut Railroad. At that time, it would reach the river port of Yakutsk and the railways would have a growing role in freight traffic across the republic (Karpov 2011).

Water transportation is used to supply goods such as oil and oil products, coal for housing and community use, wood, dry cargo, construction materials, and food. Water transportation in Yakutia is also important for passenger traffic, because the underdeveloped road infrastructure makes it the most suitable, if not the only, way to reach destinations. All lines are served by outdated 30- to 40-year-old motor ships that cannot be replaced due to lack of financial resources, In winter, these ships are repaired at Zhatai and Peledui ship repair yards. In the Soviet era, these yards built small ships and barges for Yakutia river transport.

Motor transport is also important in the republic, as it is the only means of road communication. The republic road network is 24,050 km long; of these roads, 3616 km are federal common highways and 11,218 km are regional. Winter roads (*zimniks*) constitute the major proportion (63%) of the republic motor roads. Seasonal, low-duty, and limited-capacity roads make up 90% of all motor roads (Kharitonova 2012).

The pipeline transport network in Yakutia is highly developed. The first gas-main pipeline from Tas-Tumus to Yakutsk was built in 1967. Today, there are 4047 km of operating gas pipelines and 110 km of oil pipelines at the republic level. In addition, to exploit explored reserves of the Yakutian Talakan, Srednebotucbinsk, Chayanda, Verkhnechonsk and other oil deposits, the East Siberia–Pacific Ocean (ES-PO) oil pipeline was constructed in 2009. The 1105-km ES-PO segment is functioning from the Talakan deposit to Taishet, providing oil transport to the Angarsk petrochemical plant. In Yakutia, it crosses the Lenskii, Olekminskii, Aldanskii, and Neriungrinskii districts, and has eight oil-transfer pump stations. The oil pipeline extends to Khabarovsk and Vladivostok, delivering oil and gas to the Asian-Pacific countries (Zotova 2008).

Radical economic reforms, a transition to the market economy, and a consequent sharp increase in tariff rates for domestic air travel and cargo traffic caused a deep and lingering crisis of the Russian transport system. According to the well-known Russian economist Sergei Glazyev, annual capital investment in the development of transport rapidly declined beginning in the early 1990s, and by 1999 was reduced by 60%. Currently, because of a difficult financial situation and inefficient use of technical facilities, all components of the transport system—roads, waterways, ports and technical equipment—are in a calamitous state (Egorov 2005).

The lifetimes of major equipment, buildings, ships, boats and aircraft have exceeded their prescribed limits by many times. For Yakutia with its vast territory, the most disastrous was the reduction of traffic volume by air and sea transportation (Fig. 11.2). Currently, regular air communication (once or several times per week) within the republic is only maintained between district centers. For all other settlements, a relatively stable connection is only possible in winter. In summer, travel is by cutters and boats on rivers and streams or, as in the past, by horse and reindeer. In the arctic uluses, where the average distance between neighboring settlements is around 300–500 km, the republican government finances helicopter flights. However, these flights are subject to limitations and each *nasleg* (rural settlement) has an annual limit. Moreover, a flight is not made according to a schedule but becomes possible only if it is 65% full both ways, which is very infrequent given the

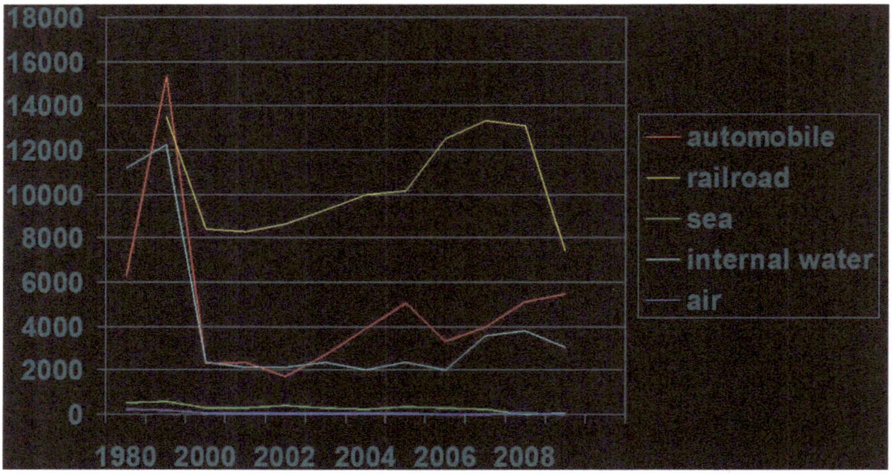

Fig. 11.2 Cargo traffic by type of public transport in Sakha Republic (Yakutia) (1980–2009)

Fig. 11.3 Winter river road (zimnik)

scarcity of population. Therefore, passengers frequently have to wait for a flight to their destination for a month or longer (Fig. 11.3).

The main type of transportation during the period of navigation is by water. Traffic is carried by Lena United River Shipping on the Lena River and, if the water is high, on some of its tributaries. Because of a current crisis of water transport in Russia, traffic on other rivers, previously plied by a shallow-draft fleet, is carried by the locals themselves on small vessels (motorboats and motor cutter boats). Passenger and cargo traffic is also carried by private individuals.

Cabotage between settlements on the coast of the Arctic Ocean has been terminated with the demise of the northern sea route system. An insignificant volume of sea shipping is carried along the Vladivostok–Zelenyj Mys line (Kolyma estuary).

Problems of Yakutia's water transport are similar to those of all Russia, i.e., decreases in the length of navigable routes and in the availability of devices to ensure navigation safety (by 1.5 times relative to 1980), and deterioration of port facilities and vessels.

As shown in Table 11.2, all transportation in Yakutia is currently via river and road. The importance of waterways for communication and of specific types of road (such as winter roads) has increased substantially, and will most likely continue in the future.

11.5 Winter Roads (*zimnik*)

Because winter roads are a characteristic feature of the transport infrastructure of northeastern Russia and, particularly, Yakutia, let us consider in some detail this type of communication.

A winter road is a type laid over the shortest distance crossing swamps, streams and other obstacles frozen in winter, and operate during that season. At present, it is the most popular type of road. Almost all rivers in Yakutia and part of the continental shelf turn into winter roads, because frozen cleared ice is convenient for the movement of even light cars. Driving on such roads is possible from November through mid-April in central Yakutia and late December through early May in northern Yakutia. Small rivers freeze more rapidly and open up earlier. There are standards and requirements for ice crossings.

According to official departmental building specifications (DBS) 137-89 "Winter Motor Roads Design, Construction, and Maintenance in Conditions of the USSR Siberia and Northeast," winter motor roads are classified as follows.

(a) **By exploitation duration**

Regular: resuming every winter for some years along the same route
Temporary: exploited for one or two winter seasons
One-time: for a single passage of a group of vehicles

(b) **By location**

Terrestrial: laid on land
Ice: laid on river, lake, water storage or sea ice
Ice crossing: across watercourses, for terrestrial winter motor roads and ordinary regular highways

(c) **By seasonal exploitation duration**

Ordinary: meant to be exploited only during periods with stable subfreezing temperatures

Table 11.2 Cargo traffic by general-purpose transport

	1980	1990	2000	2001	2002	2003	2004	2005	2006	2007	2008	2009
Transport- all including:	**18,251**	**42,947**	**15,136**	**14,930**	**14,508**	**16,179**	**17,695**	**19,493**	**19,722**	**22,578**	**23,667**	**17,598**
Automobile[a]	6332	15,356	2287	2361	1733	2701	3895	5089	3331	3995	5140	5477
Railroad[b]	…	13,500	8398	8329	8628	9282	10,002	10,176	12,546	13,310	13,098	7430
Sea	512	581	292	312	412	314	244	355	303	224	27	34
Inner water	11,199	12,213	2361	2130	2158	2375	2038	2360	2008	3568	3792	3043
Air	208	158	40	39	34	32	31	32	25	27	29	24
Pipeline	…	1139	1759	1760	1543	1476	1486	1481	1509	1455	1582	1590[c]

[a]Since 2000, cargo traffic figures have been given for large, medium, and small businesses in automobile transport, as well as for volumes carried by individual entrepreneurs (private individuals) operating commercial cargo traffic
[b]Since 2006 with the "Russian Railways" public corporation
[c]Excluding the pipeline portion belonging to the "Transneft" joint-stock company

Winter motor roads with prolonged operational lifetimes: passable in winter and partially into or entirely through summer.

Depending on effective annual traffic volume or rated traffic density, winter motor roads are divided into three categories

- Long-term (3- to 5-year) traffic volume over 100,000 net tons annually or rated traffic density normalized for a vehicle with 5-ton carrying capacity at 500 vehicles daily
- Long-term traffic volume 50,000–100,000 net tons annually or rated traffic density 150–500 vehicles daily
- Long-term traffic volume up to 50,000 net tons annually or rated traffic density up to 150 vehicles daily

- The standard requirements also presume that winter motor roads should
  - be built rapidly, of local construction materials, with maximum mechanization
  - support specified loads of all vehicle types (wheeled, crawler, sleigh) and provide for their passage at rated speeds for the required service lifetime
  - easily recover after damage caused by motor vehicles and natural effects (Departmental Building Standart #137-89 1991).

If ice thickness is 30 cm, the passage of lighter vehicles (up to 5 tons) is allowed. For ice thicker than 50 cm, cargo vehicles are allowed but for vehicles with trailers, the thickness must be no less than 70 cm. The most solid ice, according to specialists, is from March to early April. The opening and closing of crossing and movement along winter roads is determined by a special decree of the Sakha Republic government, and depends on weather conditions and the corresponding degree of river freezing.

Admissible loads and minimal distance between vehicles, dependent on ice thickness, are also regulated by the aforementioned state specifications (Table 11.3) and based on many years of technological testing.

In addition to all the technical parameters listed, winter roads, especially when laid on rivers, lakes, and gulf ice, should certainly ensure the safety of traffic, maximum possible operational duration of stream crossings, efficiency, and shortest routes between freight centers.

To accelerate ice crossing development and increase ice thickness, since the 1970s–1980s in Yakutia, various methods of artificial ice freezing have been used. The most popular and well-known is ice-building by watering with the *Grad* ("hail") sprinkling machine. With optimal water/ice balance in the mixture and at air temperature − 40 °C with weak wind, a 5- to 7-cm layer remains frozen for only 2 h, and a 25- to 30-cm layer for a day. Ice is watered down at intervals for optimal water freezing and to achieve the density and strength of natural ice. The method has been used to make crossings on major rivers, the Yakutsk–Nizhny Bestiakh in particular.

The method of accumulating ice along the crossing in early winter is very common. In this case, ice thickness is increased through repeated breakage of thin ice by river

Table 11.3 Admissible loads and minimal distance between vehicles, dependent on ice thickness

Admissible load (car or tractor weight), t	Ice thickness at average air temperature for 3 days, °C			Minimal distance between traffic lanes, m
	−10 and below	−5	0 (short thaws)	
Crawlers				
4	18	20	23	10
6	22	24	31	15
10	28	31	39	20
16	36	40	50	25
20	40	44	56	30
30	49	54	68	35
40	57	63	80	40
50	63	70	88	55
60	70	77	98	70
70	79	87	111	Single
80	88	97	123	Same
90	97	107	136	Same
100	106	118	149	Same
Wheeled vehicles				
4	22	24	31	18
6	29	32	40	20
8	34	37	48	22
10	38	42	53	25
15	48	53	60	30
20	55	60	68	35
25	60	66	75	40
30	67	74	83	45
35	72	79	90	50
40	77	85	96	55
50	82	90	114	65
60	92	100	129	75
70	103	113	144	Single
80	114	126	160	Same
90	127	139	177	Same
100	138	153	194	Same

Notes: 1. Figures for ice of freshwater rivers and lakes with greater strength than brine ice. For gulf ice, admissible loads are 20% smaller
2. Admissible load is defined for flat uncracked transparent laminar ice, frozen from lower layers. For ice with vertical veins or dry partial cracks, up to 3-cm wide admissible load is 2–30% smaller
3. During frequent thaws and for ice with wet cracks, admissible loads are half those in the table

boats with sufficiently strong boards. The postponed freezing-over and free water of early winter results in greater ice weight and therefore thicker initial ice cover.

Regular snow cleanup on rivers and lakes in early winter is also believed to be effective. However, in Yakutia this is very rare and is used only at short crossings with flat ice. Because it is relatively labor-intensive, also rare is the ice cover reinforcement method, i.e., in-freezing binding admixtures. Ice with wood fiber frozen to 5–7% of its volume is two to four times stronger than ice created by other means, for example, by water pouring over ice. In Yakutia, this method is used only at short crossings and to mend ice breaks (Faiko 1988).

At present in Yakutia, 21 ice crossings operate, with a total length of 32 km. There are 24 winter motor roads of 12,900 km total length. As a rule, those roads are laid along all major rivers (there are 3495 km of such "ice highways"), and connect almost every locality within the districts. Thus, stable communication with every settlement (even the most remote) is established only when winter roads are laid in the north (Fig. 11.4).

The longest motor roads are the Yana (Topolonoye–Tokuma–Batagai–Ust't-Kuiga–Deputatskii–Belaia Gora, 1733 km), Anabar (Mirny–Udachnyi–Olenek–saskylakh–Yuriung-Khaia, 1573 km), Arktika (Sasyr–Ugol'noe–Zyrianka–

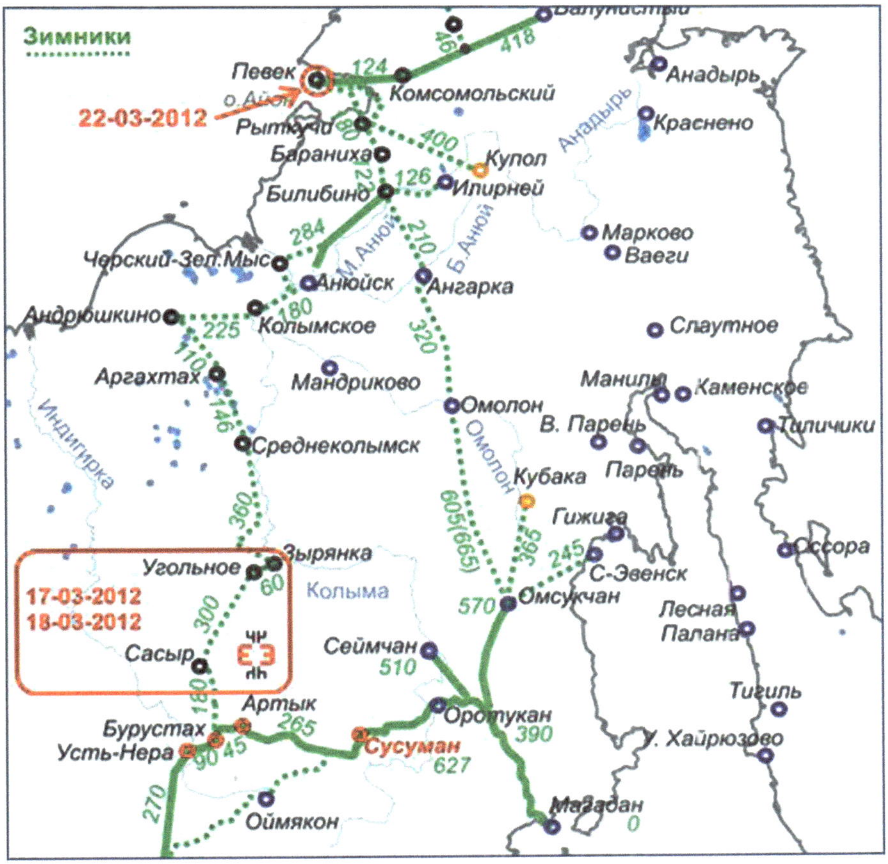

Fig. 11.4 Winter motor roads of northeastern Yakutia and Chukotka

Srednekolymsk–Andriushkino–Cherskii, 1569 km), Indigir (1123 km, 782 km of which is along the Indigirka River), and Umnas (Yakutsk–Pokrovsk–Olekminsk–Daban–Chapaevo–Turukta–Lensk, 1022 km). Some winter motor roads are extensions of year round federal highways such as the Lena, Kolyma, and Viliuy (Lebedev 2012).

11.6 Climate Change and Its Effect on Water Resources and Transport Communications of Yakutia

As pointed out by the authors of the *Climate Change and Water Resources* working paper, traffic networks are affected by changes in climate and associated events such as floods, ice and permafrost thawing, heavier precipitation, stronger winds, and less snow cover. Effects include both direct damage from, for instance, disastrous floods or road washout from rain, and those on the functioning, price, and capability of objects not designed to accommodate climate conditions predicted to dominate in the future (Baytes et al. 2008).

The transport systems of Yakutia, thus far dependent on natural resources and water ecosystems, are also under high risk of the effects of global and local climate changes.

Serious burdens are laid upon the water transport of the republic, and these are no longer far from a state of emergency. Significant damage to port buildings and equipment was caused by disastrous floods during 2001–2010. Further, in the last decade, according to observations of long-time residents, there has been more intense erosion of river banks. Given the current state of the republic's water transportation, any environmental burdens, including global climate change, will pose additional risks to the security of the people (Boyakova et al. 2010).

The aforementioned floods seriously damaged other components of the Yakutia transportation infrastructure. The 2010 flood alone led to the breakdown of seven bridges, four dams, 80 km of local roads, 11.8 km of interregional roads, and 6.5 km of seven republic roads. Interruption of traffic caused by flooding and washout inflict losses on both businesses and individuals (Departmental archives of the SR(Y) Motor Roads Department n.d.).

Climate-imposed changes in ice parameters and weather conditions have a clear influence on operation of the most popular motor roads in Yakutia, winter roads. Okumura (2009) indicates two regional winter road types. These two are those traversing predominantly water surfaces and therefore dependent on the conditions of river (or sea and lake) ice, and those on land and therefore very sensitive to alterations of permafrost and snow cover. Global warming in the future may change the times of ice freezing and melting on the above three water bodies, imposing adjustments of the work and life schedules of the people (Okumura 2009).

Table 11.4 presents the opening and closing dates of the main Lower Bestekh–Yakutsk winter road, which has ensured the delivery of goods to the capital of the republic and its northern and Vilyuy districts over the last 10 years.

As seen in the Table 11.4, the situation has been relatively stable, owing to peculiarities in the freezing over of the Lena. Specifically, the river's average ice thickness and ice cover are respectively much greater and more extensive relative to the rivers of European Russia and lakes. Apparently, a longer chronological retrospective of observations after the establishment of ice roads on the Lena is required to draw more concrete conclusions on the change in duration of ice cover.

The greatest risk is posed to small rivers and streams which, in contrast to metropolitan and central crossings, lack special posts and have less regulated safety. There are also a large number of ice crossings that are built by people unauthorized to do so Annually in the vicinity of Yakutsk, 12–17 such crossings are spontaneously organized by people without authorization over the Lena alone. With climate change, observations over many years of freezing and opening up of water bodies, as well as evidence from folk meteorology, do not apply to the present reality. Consequently, much stricter control is necessary over activities at crossings and winter roads.

Change in the water regime (later freezing and earlier opening up) seriously alters the traditional economic activities of indigenous peoples. For example, many hunters in northern Yakutia are complaining that belated establishment of the winter sledge road disrupts their natural economic cycle, worked out over centuries. Because of a later freezing of swamps and streams and wider spread of phenomena such as *taryns* (water flowing over ice), the hunters are forced to delay their season up to a month, which inevitably affects their spoils and thereby their profits. Over recent years, there has been an increase in the number of accidents connected with reduced safety of movement. This in turn is a result of changes in ice characteristics and weather conditions.

Permafrost thaw also has a negative impact on Yakutia communication lines. Because ~90% of permanent, seasonal, and local roads are unpaved, the people have become concerned with more frequent collapses of road segments, especially in zones where what are known as *alas* systems prevail (Fig. 11.5). Soil deformation from permafrost thawing and washout from increased precipitation lead to breakage of traffic-bearing surfaces and other components of the transportation infrastructure, such as buildings, ports, berths, and pipelines.

In association with the aforementioned phenomena, climate change exerts strong influences on the state and development of the transportation infrastructure, and therefore on the economy and well-being of the people. Among these influences, we highlight the following.

- Destruction of roads, buildings, pipelines and other constituents of the transport infrastructure by the thawing of frozen ground
- Change of transport schemes as a result of later freezing and earlier opening up of rivers

Table 11.4 Yakutsk–Lower Bestekh winter road opening and closing dates

	2001/2002	2004/2005	2005/2006	2006/2007	2008/2009	2009/2010	2010/2011
Opening	15.01.2002	14.12.2004	20.01.2006	05.01.2007	05.01.2009	15.01.2010	05.01.2011
Closing	15.04.2002	05.04.2005	15.04.2006	15.04.2007	15.04.2009	15.04.2010	15.04.2011

Source: Departmental archive of the SR(Y) Motor Roads Department

Fig. 11.5 Destruction of roads by thawing of frozen ground. Megino-Kangalasskii ulus (district), Suola village, 2010

- Increased spending on prevention and reduction of the consequences of natural disasters (damaged roads, bridges, port buildings and other parts of the transport infrastructure)
- Increased mortality rate among the population because of an increase in accidents associated with reduced safety of movement caused by changes in ice characteristics and weather conditions.

11.7 Conclusion

The history of the Arctic peoples has many examples of successful adaptation to the impacts of negative phenomena connected with weather and climate, especially floods and drought. Nevertheless, additional measures of adaptation at regional and local levels are required to diminish the adverse effects of climate change, which are growing at alarming speed in the high latitudes when compared with other regions of the earth. These measures could reduce the vulnerability of societies to climatic changes, in both the short and long run. It is particularly important to elaborate various strategies of adaptation at the governmental level for sectors such as transport, energy economics, water management, agriculture, and health care. Regrettably, in Yakutia there remain serious obstacles of financial, technological, political, social, institutional, and cultural character, which restrict the possibilities of adaptive measures and their efficiency.

To diminish the negative consequences of climate change on the social and economic development of the Sakha Republic, it is necessary to develop a complex of measures directed at the prevention or alleviation of all the aforementioned problems.

References

Baytes BK, Kuntsevich ZV, Paliutikoff SU, Paliutikoff ZP (eds) (2008) Climate change and water resources. Technical paper of Intergovernmental group of experts on climate change. IGECC Secretariat, Geneva

Boyakova SI (2001) Osvoenie Arktiki i narody Severo-vostoka Asii (XIX vek – 1917). [The development of the Arctic and the peoples of North-east Asia (the nineteenth century – 1917)]. Nauka, Novosibirsk. (in Russian)

Boyakova SI, Vinokurova LI, Ignatieva VB, Filippova VV (2010) Yakutia v usloviyakh global'nikh klimaticheskikh izmenenii: uyazvimost, ryski, socialnaya adaptaciya [Yakutia in the context of global climate change: vulnerability, risk, social adaptation]. Severo-vostochnii gumanitarnii vestnik 1:22–25. (in Russian)

Departmental archive of the SR(Y) Motor Roads Department (n.d.)

Departmental Building Standart (DBS) #137-89 (1991) Proektirovanie, stroitel'stvo i soderzhaniye zimnikh avtomobil'nikh dorog v usloviyakh Sybiry i Severo-vostoka SSSR [Winter motor roads design, construction, and maintenance in conditions of the USSR Siberia and Northeast]. Moscow (in Russian)

Drachev PT, Knoll VA, Nikitin AN (2005) Proekt Noosferniye thansportniye sistemy Sibiri i Dal'nego vostoka: itogi relizacii [No-sphere transport systems of Siberia and Far East: realization sum up]. Novosibirsk, pp 422–423(in Russian)

Egorov EG (ed) (2005) Yakutia: razmesheniye proizvoditel'nykh sil [Yakutia: placement of the productive forces]. Nauka, Novosibirsk. (in Russian)

Faiko LI (1988) Ledyanie mosty [Ice bridges]. Yakutskii Knizhnyi Dom, Yakutsk. (in Russian)

Karpov Y (2011) Most cherez Lenu: plusy i minusy [Bridge over Lena: pros and cons]. Gazeta Yakutiya 2011:3. Yakutsk (in Russian)

Kharitonova T (2012) Ministr transporta Yakutii Semen Vinokurov: "Problem, k sozhaleniyu, nemalo" [Transportation Secretary of Yakutia Semen Vinokurov: "Unfortunately, problems are many"]. Gazeta Zhizn' Yakutska 2012:8–9. Yakutsk (in Russian)

Khomyakov P, Kuznecov VI, Alferov AM et al (2005) Vliyaniye global'nikh izmenenii klimata na funkcionirovaniye economiki i zdorovye naseleniya Rossii [The impact of global climate change on the functioning of the economy and the health of the population of Russia]. Lenand, Moscow. (in Russian)

Lebedev A (2012) Zimnya doroga – samaya vernaya [Winter road is the right one]. Gazeta Nashe Vremia 48:13. Yakutsk (in Russian)

Okumura M (2009) Siberian winter roads on the frozen river and ponds, as public goods depending on the local environment. International Workshop "Regional public sphere and environment in Slavic Eurasian and Japan". February 28–March 1, p 5. Sapporo

Zotova E (2008) Pervaya neft'[First oil]. Gazeta Vedomosti 2008. Yakutsk. (in Russian)

Chapter 12
Adaptation Strategies for Risk and Uncertainty: The Role of an Interdisciplinary Approach Including Natural and Human Sciences

Makoto Okumura

Abstract The human and social response in Eastern Siberia to possible changes induced by global warming is the main theme of the Research Institute for Humanities and Nature (RIHN) Siberia Project. This presentation explains why we must use an interdisciplinary approach including natural science and human sciences to tackle this theme.

The behavior of local society and ordinary people, especially indigenous peoples, is not always based on knowledge of modern natural science, but rather on local empirical knowledge or social norms. The latter may have been collected, conceptualized, and investigated by human and social scientists. Such empirical knowledge and social rules have rarely been the product of theoretical thinking or optimal design, but of an accumulation of tacit knowledge inductively attained and tested through interactions with other people in one society, among different societies, and the natural environment, in a trial-and-error manner over time. If the effectiveness of such knowledge and rules has been tested only within the range of past environmental change, then we cannot be certain of their effectiveness and applicability in the future, especially in relation to possible climate changes associated with global warming or to major changes in the social environment (for example, in demography). In other words, empirical knowledge can be interpolated but not extrapolated.

In comparison, natural sciences such as mathematics and physics have a wider range of applicability and extrapolatability. Even in novel case settings, we can simulate possible situations. When we execute a simulation, however, we must determine the range and step of time and space. At such times, natural scientists usually want to establish consistency and fit with empirically observed data. As a result, they pay attention only to the interpolation of known phenomena and exclude the possibility of never-experienced or unexpected phenomena.

M. Okumura (✉)
International Research Institute of Disaster Science (IRIDeS), Tohoku University, Sendai, Japan
e-mail: mokmr@m.tohoku.ac.jp

T. Hiyama, H. Takakura (eds.), *Global Warming and Human - Nature Dimension in Northern Eurasia*, Global Environmental Studies,
DOI 10.1007/978-981-10-4648-3_12

To enhance a society's capacity to adapt, we must encourage all scientists to take the risk of extrapolation. Natural scientists need not propose an accurate expectation in a strictly limited setting, but should clarify the range of oscillation of nature and indicate the possibility of new phenomena appearing outside it. Human and social scientists should investigate whether local knowledge includes any insight into the appearance of new phenomena.

Keywords Risk • Uncertainty • Adaptation • Interdisciplinary approach

12.1 Risk, Uncertainty and Ignorance

The word "risk" may be defined in a vague or broad sense, such as "some negative problem, which may occur sometime in the future, but we cannot surely know, when, where, and how strong it will be". To discuss the role of scientific knowledge in such risk, we should more carefully define its underlying concept.

Frank Knight (1885–1972) was a Professor of Economics, Chicago University, and one of the world's leading economists. He made significant contributions to many problems of both economic theory and social philosophy, and carefully distinguished economic risk and uncertainty in his famous book *Risk Uncertainty and Profit* (1921), based on his Ph.D. dissertation at Cornell University. He defined risk as a situation in which outcomes are unknown but the governing probability distribution is known, and decision-making rules such as expected utility maximization can be applied to such a situation. In contrast, an "uncertain" situation is defined as one in which outcomes can be listed beforehand but probability cannot be measured. Knight argued that uncertainty gives rise to economic profits that perfect competition cannot eliminate. In this line of thinking, we can also define a "perfect unknown" or "ignorance" as a situation in which we cannot anticipate the existence of that outcome until its occurrence.

Risk management is the identification, assessment, and prioritization of risks (defined in ISO 31000 as the effect of uncertainty on objectives, whether positive or negative), followed by coordinated and economical application of resources to minimize, monitor, and control the probability and/or impact of unfortunate events or maximize the realization of opportunities (International Organization for Standardization 2009). In risk management, risk is assessed by the multiplication of impact and probability. This usually dismisses the control of low-probability events and uncertain phenomena, including events of considerable negative impact (Hubbard 2009). Such neglect results in poor response following such events and blame from the public, such as the seawall break in the town of Taro in Miyako City caused by the 11 March 2011 tsunami and Fukushima Nuclear Power Plant disaster. The natural science approach usually explains phenomena through forward analysis through time. If there is any uncertain point in the process, it cannot give reliable probabilities for the following situations. Natural science also uses a model with parameters that are sometimes adjusted to fit measured and observed phenomena.

As a result, the model is tuned to reproduce only "natural" and "ordinary" phenomena. It fails to produce "unnatural" and "extraordinary" situations, even if the model logic has the potential to do so. Compared with natural science, human and social sciences take a more flexible approach (including reverse analysis), and are not confined to capturing phenomena quantitatively. Based on case studies, historical research and interregional community comparisons, these sciences can discover phenomena that have not been quantitatively captured by natural science.

12.2 Adaptation Concept in Disaster Management and Exceedance Hazard Problem

12.2.1 Adaptation Strategies in Disaster Management

In the field of disaster management, factors to determine damage and loss in a disaster are conceptualized in Fig. 12.1, which shows time-dependent change of socioeconomic activity level. At some point in time, a natural hazard arises in a region, and socioeconomic resources and facilities are exposed to this hazard. If the strength of the social system is inadequate to resist the magnitude of the hazard, damage will ensue immediately. This instant damage can be captured by the multiplication of hazard magnitude, exposure, and vulnerability. Disaster response and recovery activities follow. Then, socioeconomic activities recover and at some point, their level reaches that prior to the disaster. The speed of this process was called "resilience" by Bruneau (2003). Until the full recovery date, economic activity is somewhat suppressed. Bruneau also proposed a "Resilience Triangle," which gauges total loss from the disaster in Tierney and Bruneau (2007). Of course, resilience may be less for disasters with greater damage, so it is difficult to quantitatively separate resilience from vulnerability.

Based on the concept of hazard, exposure, vulnerability, and resilience, we can classify adaptation strategies into several groups. Because we cannot directly control natural hazards, there are three ways to reduce total loss, i.e., decrease exposure, reduce vulnerability, and increase resilience. The first, decreasing exposure, is called risk aversion and includes land-use control that avoids the use of risk-prone areas or temporal evacuation accelerated with disaster alert This requires substantial

Fig. 12.1 Damage and loss by natural hazard

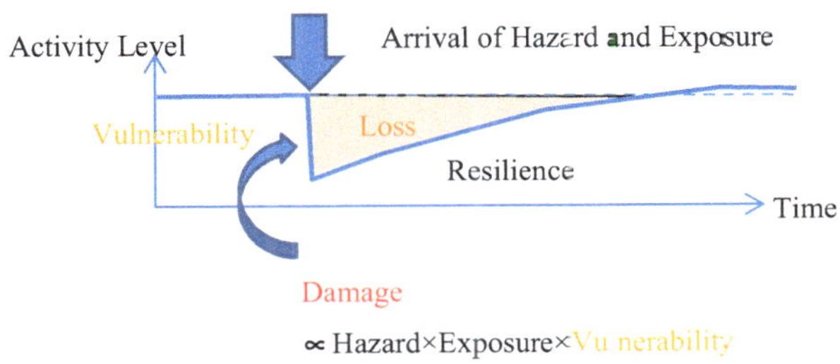

social cost for relocation and lifestyle change, and sometimes violates traditional ways of life such as hunting, fishing, and gathering. The second strategy, reducing vulnerability, is preventative. A good example is building a dam or riverbank against flood risk. This usually requires technological costs such as funds, materials, and land for facility buildings, but does not require change of lifestyle. The third strategy, increasing resilience, is considered mitigation. It includes preparation of backup or spare facilities, buying insurance, or other financial systems.

Until the Great East Japan Earthquake on 11 March 2011, the Japanese government mainly attended to the second strategy of disaster prevention. This included the policy to blockade, isolate, and weaken the hazard by physically resistant facilities or building facilities stronger and more durable. In the design of such disaster prevention facilities, one must set a certain target hazard level to determine the material, size, location, and mechanical strength of the facility. However, this design process cannot certify the performance for hazards greater than the design target. In other words, we cannot avoid problems of hazard exceedance or unexpected hazards in any disaster prevention policy.

12.2.2 Vulnerability for Hazard Exceedance or Unexpected Hazards

For simplicity of discussion, we assume here that the exposed facility maintains perfect resistance to hazards no higher than the design level and there will be no damage. Damage to a fully exposed facility is given as the multiplication of hazard and vulnerability, and the latter can be calculated as damage divided by hazard.

According to damage in the case of hazard exceedance, we can classify fragility and vulnerability into three types (Fig. 12.2), for which damage is plotted for varying hazard magnitudes. A Type 1 facility retains a certain amount of damage reduc-

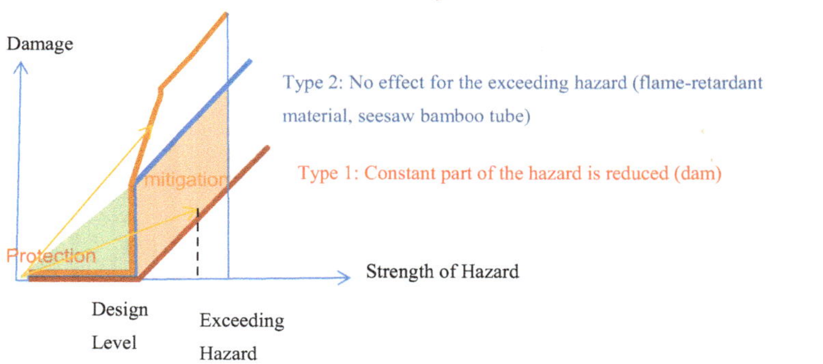

Fig. 12.2 Three types of fragility curves

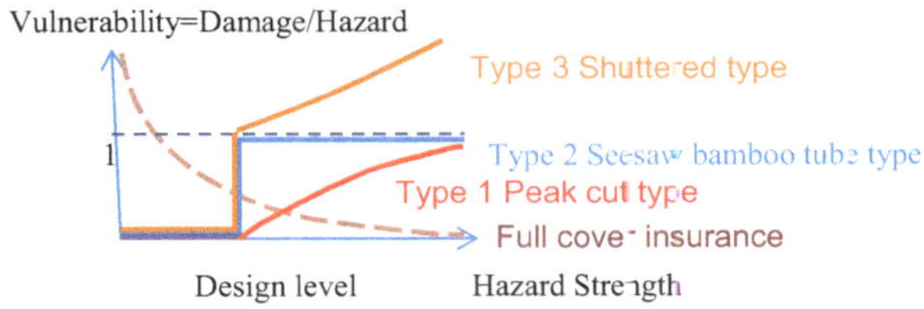

Fig. 12.3 Three types of vulnerability

tion even with hazard exceedance. A dam with a certain water volume for flood control is classified into this category. A Type 2 facility loses its damage reduction effect after hazard exceedance. This category includes flame-retardant material of buildings against fire risk. This material does not catch fire for a certain heating duration, but once it does ignite it will burn thoroughly, as with typical building material. Type 3 facilities sustain greater damage under hazard exceedance. Coastal wood against tsunami hazards is one example. It somewhat weakens the strength of tsunami wave. However, with a powerful tsunami, trees are uprooted and float away, striking buildings distant from the coast, increasing the damage there. Another example is a high seawall overtopped by the great 2011 tsunami. Because of such seawalls, some local people were overconfident of their safety and did not evacuate promptly. Therefore, the tsunami destroyed the seawall and killed people who would have otherwise evacuated without such seawalls. Many disaster prevention facilities may be categorized as Type 3 in the face of great exceedance hazards such the 2011 tsunami.

Figure 12.3 shows the vulnerability of facilities categorized into the three types. This is based on vulnerability defined as damage divided by hazard, and the damage scale is standardized by the amount of damage if the facility were not there. From the figure, vulnerability for Types 2 and 3 dramatically increase once hazard strength exceeds the design level, while vulnerability gradually increases for a Type 1 facility. We also plotted the vulnerability of full-coverage insurance, which gradually decreases upon hazard exceedance.

12.2.3 Use to the Design Limit and Sudden Destruction

Type 2 and 3 facilities have the possibility of sudden destruction. This characteristic entails substantial damage if many people attempt to use facilities up to their design limits. Unfortunately, such use is not an exception but a frequent behavior based on rational decision-making. Generally, we feel scarcity premium for few supplied goods and services. If many suppliers are sensitive to the risk of destruction and hesitate to provide something for use of a facility near its limit, risk-taking use may have scarcity premium and economic incentive. As a result, a risk-taking individual tries to use the facility up to its limit. However, we cannot observe the hazard level perfectly or the actual strength of the subject facility, so some of them will be at risk of sudden destruction or collapse.

There are two approaches to avoid sudden destruction. One is to increase the accuracy of destruction forecasting and monitoring, and inform people of the limits of use more accurately. This approach sometimes incentivizes financial support to scientific research on natural hazard phenomena, but requires a certain sacrifice. The other approach is to build a social system that discourages use of facilities near their limits. To establish a more modest limit and prohibit use during limit exceedance is one possibility. However, owing to the economic incentive for scarcity, violations of such restrictions can easily occur. It is also difficult to properly determine the limit. To require extra cost for use in limit situations appears more effective. Extra charges for facility usage is a possible policy option.

12.3 Possible Effect of Global Warming in the Far North

12.3.1 Winter Road System in Siberia

Siberia spreads across 2500 km from the Arctic Ocean to its southern border with China and Mongolia, and about 8000 km from the Ural Mountains to Bering Strait. The density of the transportation system is very low. In winter, the temperature falls to −50 °C and water transportation becomes impossible because of the freezing of major rivers. However, ice with a minimum thickness of 1 m can support a vehicle. Thus, the *avtozimniki* (winter road in Russian) is built atop the frozen rivers. This is shown by dotted lines in Fig. 12.4, which is a transport network map of the Sakha Republic.

There are 22,000 km of national highways in the Sakha Republic, two thirds of which are winter roads. Around Yakutsk, a winter road is used for about 6 months, from early November through late April. Asphalt roads in Sakha extend only 623 km, 3% of the total highway length. The asphalt becomes soft in summer, when the temperature exceeds 30 °C. Heavy traffic kicks out macadam in the pavement and makes crevasses and cracks in it. These cracks widen in winter by the pressure of freezing water, and new cracks appear. Therefore, asphalt pavement requires substantial maintenance as well as construction cost. Recently, a vinyl sheet has been inserted between the pavement and roadbed to prevent penetration by water, but this cannot be widely used because of cost issues. The remaining 6900 km is unpaved earthen road. This becomes rough in summer when the earth layer (~1.5-m-deep) atop the permafrost melts to become mud.

In Fig. 12.4, solid lines are ordinary roads with year-round use and gray lines are navigable rivers. Circles are capital cities of local municipalities, many of which are only reachable in winter from Yakutsk, the republic capital, via the winter roads. Even on the all-season roads in the figure, there are no bridges crossing major rivers, so a ferry service is required. For example, there are no bridges over the Lena River, so the winter road provides much easier crossing. In winter, snow on the ice surface is removed to make it smooth and stimulate freezing through cooling from the cold air.

Fig. 12.4 Transport network of Sakha Republic in East Siberia

12.3.2 Use of Winter Roads

As mentioned above, in the Sakha Republic, year-round service of ordinal road transportation is difficult and the winter road system has a major role. Before the collapse of the Soviet Union in 1991, foods and other necessities were airlifted by helicopters to the settlements and villages, under the responsibility of the federal government. Such costly means became impossible after the collapse, thus the existence of small rural villages now strongly depends on the winter road system.

As stated above, the winter road system around Yakutsk is available for 6 months, from November through April. However, for particular purposes, the transportation season is shorter. For example, maintenance of electric power transmission lines in rural areas must be scheduled in winter, considering the transportation of materials and construction vehicles. Furthermore, they cannot schedule field work between December and February, when the temperature drops to less than −40 °C. This is because of the threat to life of workers in case of vehicle breakdown in rural areas. Therefore, the electric power company *Sakha Power* plans all maintenance work for the months of November, March, and April.

The winter road system has been used based on empirical and local considerations, rather than on scientific knowledge. Officially, a commission composed of specialists from the Department of Transportation, Police, Road Construction and Water Navigation, as well as the Emergency Department, determines permitted dates of use. This is based on whether the measured ice thickness is greater than 1 m. The official opening date for vehicles under 1 ton is fixed on 20 December, and for heavier vehicles on 1 January. A 2- or 3-day suspension is sometimes ordered. However, many vehicles begin to use the roads much earlier than those dates. In spring, there are many users in early May, after the formal ending date in April, resulting in many vehicles falling through the ice.

Ice road collapse occurs suddenly and when it does, there is serious damage to vehicles and goods, and injury to people. From the typology described in Sect. 12. 2, an ice road system, especially one crossing deep rivers, is categorized with Type 3 vulnerability. Therefore, a social system to discourage its use to the limit should be forged. For example, setting a toll for the risky spring season should be implemented.

12.3.3 Global Warming in Siberia

Siberia is more strongly affected than other parts of the world by global warming. For example, the rise of the average temperature from 1975 to 2000 was 0.5 °C greater than the global average, and 2.5 °C in Yakutsk. Expectation of average temperature rise in Siberia in the fourth report of the Intergovernmental Panel on Climate Change was 1.5 °C in 25 years, with the increase over a century rise at 5–6 °C. Winter precipitation is believed to be on the increase, with decreases in summer. Rivers may still freeze under such climate change, but its timing and that of melting may change greatly.

As stated above, there are several works dependent on the winter road system that can be done only during limited periods in the severe Siberian climate. Future global warming may alter the timing of ice freezing and melting on rivers, thereby mandating adjustments to the work and life schedules of the people. Local empirical knowledge regarding the winter roads will lose its applicability soon. We must provide alternative knowledge and technology based on more explicit and scientific knowledge.

We have suggested some technologies to be developed for the winter road system in the age of global warming. First is the important measurement of ice thickness. Radar measurement technology is promising for short-term alerting or limiting the passage of heavy vehicles, because it is connected with the mechanical assessment of ice strength. Second is mid-term forecasting of ice thickness decrease and collapse, based on upper streamflow conditions and remote sensing information of upper basin snow thickness and melt. These technologies are too military-related to be operated by foreign researchers, so Russian researchers should advance such studies. However, such research may be a double-edged sword, because it encourages usage to the limit. Social policy to discourage such usage should be investigated in parallel.

12.4 Limit of Local Knowledge Under Regime Change of Natural Ecosystems

12.4.1 Homeostasis in Ecosystem

From the perspective of systematic biology, every ecosystem has internal self-sustaining feedback mechanisms to return to the prior state after external shocks or changes of boundary conditions. Clearly, if external shocks or perturbations exceed certain limits, such restoring forces cannot be anticipated. As shown in Fig. 12.5, systems transition to different equilibria. We can call the range of the state that converges to one equilibrium a "regime." Under a stronger shock or perturbation, the state sometimes transforms into a regime different from the previous one.

12.4.2 Limit of Local Knowledge

Local society and ordinary people, especially indigenous ones, do not always determine their behavior based on the knowledge of modern natural science, but rather on local empirical knowledge or societal norms. The latter have been collected, conceptualized, and investigated in the human and social sciences. This empirical knowledge and rules have rarely been the output of theoretical thinking or optimal design. Instead, they are an accumulation of tacit knowledge, inductively attained and tested through interactions with the natural environment in a trial-and-error manner over long history. The local knowledge and rules have been tested for effectiveness only for changes in environment, usually in the present regime. We cannot assure effectiveness and applicability of this knowledge in the future, following

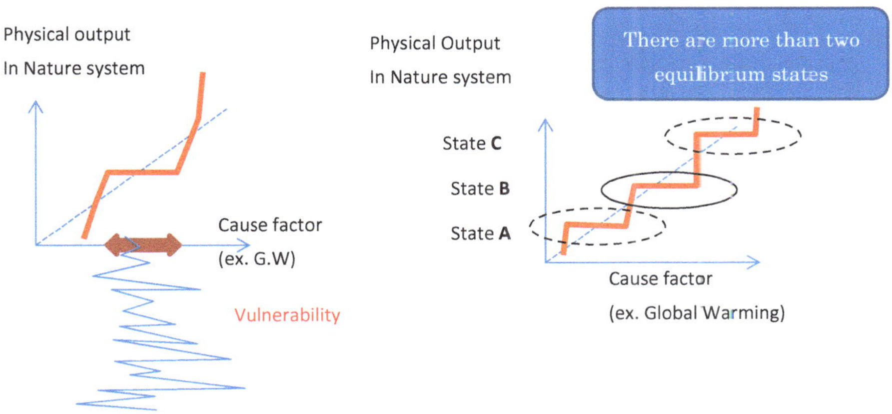

Fig. 12.5 Homeostasis in eco-system and several regimes

transition to another regime caused by climate change from global warming or by major changes in the social environment, such as demographics. In other words, empirical knowledge can be interpolated within the present regime but not extrapolated beyond it.

To go beyond the regime, comparative studies appear to be important for human and social scientists. They may provide knowledge about the situation under different regimes.

12.4.3 Beyond "Scientific" Scientists

Compared with empirical knowledge, natural science theory such as mathematics and physics have a wider range of applicability and are more extrapolatable. Even for a never-experienced case setting, we can simulate the potential situation. However, when we run a simulation, we must determine the range and step of time and space. At such a time, natural scientists usually seek consistency and the fitting of observed data. As a result, they consider only the interpolation of experienced phenomena, and exclude the possibility of never-experienced or unexpected phenomena.

To investigate the adaptation limits of society, we must encourage all scientists to risk extrapolation. Natural scientists need not promise accuracy in a strictly limited setting, but should clarify the potential range of variation of nature and new phenomena. Human and social scientists should find similar samples of people and community from any location on the globe, which encourage such extrapolations of natural scientists.

References

Bruneau M (2003) A framework to quantitatively assess and enhance the seismic resilience of communities. Earthq Spectra 19(4):733–752

Hubbard D (2009) The failure of risk management: why It's broken and how to fix it. John Wiley & Sons Ltd., Hoboken

International Organization for Standardization (2009) Risk management — principles and guidelines on implementation. ISO/DIS 31000, Geneva

Knight F (1921) Risk, uncertainty and profit, ISBN 978–0–9840614-2-6

Tierney K, Bruneau M (2007) Conceptualizing and measuring resilience: a key to disaster loss reduction. In: TR News, pp 14–17

Chapter 13
Negotiating Risk and Responsibility: Political Economy of Flood Protection Management in Northern Finland

Monica Tennberg, Terhi Vuojala-Magga, Joonas Vola, Heidi Sinevaara-Niskanen, and Minna Turunen

Abstract Floods occur every spring in Northern Finland. These floods attract locals to the shores of rivers to admire the movement of melting ice and strength of the water. Floods also damage buildings, roads and disturb everyday life in many ways. The EU Flood Directive (2007) and its national implementation recently require local and regional authorities as well as inhabitants to take the threat of floods more seriously and prepare better for them than before. This chapter investigates the rationalities in developing regional flood management plans for two major flooding rivers in Lapland, for Ivalo and Kemi Rivers in 2013–2015. Although, there were some similarities in debates concerning the flood management planning in both cases, outcomes of the two participatory planning processes were quite different in terms of assessing the level of risk and defining responsibilities to tackle it. Regional flood management is politically much larger issue than to find a feasible and economically sensible technical solution for effective flood protection.

Keywords Flood Protection • Lapland • Finland • Responsibilisation • Risk • Neoliberalism

13.1 Introduction

In this chapter, the focus is on the development of flood management plans in Northern Finland concerning two nationally recognized flood risk areas along the Ivalo and Kemi rivers. Over 2013–2015, motivated by the EU Flood Directive (2007)

M. Tennberg (✉) • T. Vuojala-Magga • J. Vola • M. Turunen
Arctic Centre, University of Lapland, P.O. Box 122, 96101 Rovaniemi, Finland
e-mail: monica.tennberg@ulapland.fi

H. Sinevaara-Niskanen
Gender Studies Unit, Faculty of Education, University of Lapland,
P.O. Box 122, 96101 Rovaniemi, Finland

T. Hiyama, H. Takakura (eds.), *Global Warming and Human - Nature Dimension in Northern Eurasia*, Global Environmental Studies,
DOI 10.1007/978-981-10-4648-3_13

Fig. 13.1 Northern Finland in the Arctic region

and its national implementation, plans to improve flood protection in Northern Finland were developed through a participatory process, including assessment and mapping of flood risks and identification of alternative flood protection measures and their effectiveness, toward making regional flood management plans (see Fig. 13.1). These processes resulted in a re-division of risk and responsibility for flood protection among local residents, municipal, regional and state authorities, and other stakeholders. These two cases show very different solutions to regional flood management, but tell a common story of negotiating risk and responsibility over flood protection. The chapter summarizes findings from interviews, analyses of media material, official documents and background expert material for flood protection, and on a participatory method developed by following public hearings and meetings organized in connection with flood protection planning in Ivalo and Rovaniemi.

13.2 Flood Protection in Finland

Since the mid-2000s, Finland has taken important steps in flood protection as a member of the European Union (EU). National attention to flood protection measures is required by the EU Flood Directive (2007), including preparation of flood risk assessments and maps, identification of effective flood protection measures, and development of flood management plans. These measures have to be coordinated with another, earlier EU directive, the Water Directive from (2000). That directive applies to all kinds of floods (rivers, lakes, flash floods, urban floods, and coastal floods including storm surges and tsunamis). The directive includes a time table: By 2011, states were to undertake a preliminary flood risk assessment and identify major flood prone areas, by 2013 to develop flood hazard maps and flood risk maps for those areas, and by 2015 flood risk management plans must have been finalized for the areas. The flood management plans should address: (1) flood prevention by avoiding construction of houses and industries in flood-prone areas; (2) protection through measures to reduce the likelihood of floods and/or flood impacts in a specific location; and (3) general preparedness by informing the public about flood protection issues. The flood management plans will be made for 6 years at a time, and the first round of these plans cover 2016–2020 in Finland. The regional approach in flood protection management aims at enabling greater attention to local conditions in the planning of flood protection measures. The Ministry of Agriculture and Forestry approved the submitted plans in December 2015, except the one concerning the Kemi River, which was accepted only partially (MMM 2015).

Nationally, several government working groups have discussed flood risks, their assessment, and protective measures in their reports (Suurtulvatyöryhmä 2003; Tulvavahinkotyöryhmä 2006; Tulvariskityöryhmä 2009). Also, some measures have been implemented at regional and local levels. For example, regional climate centers assessed flood risks on waterways and municipalities for city floods in 2009–2011. Flood maps available to the general public were published in 2013. Nationally important flood risk areas were listed, including the two rivers in question here (the Ivalo and Kemi). Major changes in national legislation followed (Law on Flood Risk Management 2010), including a new approach to flood compensation (came into force 2014). This meant the abolishment of state compensation for flood damages, which used to cover up to 80% of those damages caused by a flood occurring approximately once in every 20 years. The aim of the national flood management legislation was to reduce flood risk and damage as well as improve preparedness for floods in the general framework of sustainable use of water resources and needs for protection. Also, new instructions for buildings in inland and coastal areas were issued in 2014 (Parjanne and Huojuna 2014). These instructions concern house and property owners as well as municipalities. Municipalities are important in flood protection management because they have a monopoly in city planning in Finland and are responsible for ensuring that new properties will not be built in areas threatened by floods. A widely used aim in city planning is to take into consideration "very rare floods" (occurring every 100–200 years). This means that for inland areas, the recommendation is to allow new properties be built with consideration of the once-every-100-year level (available from flood maps), with some flexibility depending on property type and environmental conditions (Parjanne and Huojuna 2014).

Since 2014, property owners have been required to buy insurance against floods, and such insurance is part of property insurance offered by private insurance companies. Minimum flood insurance is included in most cases within the compulsory real estate insurance but, for better coverage and compensation, special insurance is required. To have such private insurance in force, the insurance companies require that the property be located in a non-flood-prone location and that flood protection measures are in place. This is the responsibility of the property owner. According to the new law, only floods that are considered "rare", occurring once every 50 years, are covered by the new insurance system. This estimation of flood risk level is done by regional environmental centers, which are responsible for flood risk assessments, mapping, observation, and warning systems. Flood insurance is included in most products offered by Finnish insurance companies, but depends on the extent of the contract. There are differences in coverage extent, and damage to yards or property outside houses is not always covered. The deductible differs greatly between companies (Finanssialan keskusliitto 2014).

The above change in the Finnish flood compensation system necessitates renegotiation of the responsibilities for flood protection measures. Local and regional flood protection planning started in 2013 in various parts of the country, including the Ivalo and Kemi River areas. These plans were prepared by "flood groups," which consist of representatives from regional councils and municipalities, and are supported by experts and their background reports. Both regional flood management plans were accepted at the end of 2015, but the one concerning the Kemi River and Rovaniemi areas was accepted only conditionally. Final approval of the management plan depends on progress in regional planning toward implementing environmental impact assessments and other administrative procedures related to the flood reservoir plan (MMM 2015). The two cases from northern Finland show different interpretations of flood risk and identification of means to deal with risk and responsibility for flood protection. However, the common element in both cases is the privatization of risk, either to property owners or other actors, such as hydropower companies or, in the case of Kemi River, combinations of the two.

Concern over climate change and its impacts on the flood situation has not been the driving force in Lapland flood management. Current estimation is that in the short-term (20–30 years), the threat of floods may grow somewhat, but in the long-term (by 2080) the risk will decrease (Veijalainen et al. 2012). However, the climate argument has been used in the local and regional debates concerning flood protection measures, especially to support the development of renewable energy use, in particular hydropower, by building a new reservoir and power plant in Lapland.

This chapter is based on our long-term studies of extreme weather events, such as floods and storms, in the context of climate change adaptation. These studies were part of the CAVIAR, Community adaptation and vulnerability in Arctic regions (2007–2009), MISTRA Arctic Futures (2012–2014) and CAVIAR2 (2012–2014) projects. The material for this work was collected through anthropological field work and interviews in Ivalo by Terhi Vuojala-Magga, supported by our natural science expert on flood information, Minna Turunen (Tennberg et al. 2010; Vuojala-Magga and Turunen 2013). Also used were data from interviews and participatory research by political scientists in Rovaniemi, by Monica Tennberg and Joonas Vola, plus conceptual development of neoliberal responsibilization in collaboration with Heidi Sinevaara-Niskanen (Sinevaara-Niskanen and Tennberg 2012) and Joonas

Vola (Tennberg and Vola 2014). For these papers, eight semi-structured interviews were conducted in Rovaniemi during 2014–2015 among stakeholders involved with flood protection. These included representatives from the city of Rovaniemi, Regional Council of Lapland, environmentalists, a local residents' association, and a hydropower company. Two meetings followed, organized by the city for residents. In Ivalo, 29 interviews were conducted, 22 in 2007 and seven in 2012. Public meetings then followed. Regional and local newspapers, *Lapin Kansa* and *Inarilainen*, were perused to track the public debate about local flood protection.

13.3 Neoliberal Flood Protection

Adaptation means adjustment of societal practices in response to environmental and other changes (IPCC 2014). In practice, much of the adaptation takes place locally. Studies show that with modest national governance of adaptation, local and regional practices have great diversity (Keskitalo et al. 2011). The municipalities along the two rivers, Ivalo village and the city of Rovaniemi, are *units of local self-government* whose autonomy is traditionally very strong in Finland, with their own decision-making and financial powers, and responsibility for city planning. Municipalities are important for making policies and strategies that are sensitive to local contexts and community needs. Ivalo and Rovaniemi are also communities with a limited space for interactions, identities, and institutions (Rannikko 1997; Raitio and Rannikko 2006). Municipalities are also *regulatory communities* where social interaction takes place, whereas communities are localities with a combination of people, resources and practices within or beyond the limits of a given municipality, for example through infrastructure (Agrawal 2005). Effective flood protection requires close cooperation between local residents, local and regional authorities, and other stakeholders. In case of a major flood, communities are important for local flood protection and rescue measures, including communication and self-help during extreme events and associated problems, as the case of the Ivalo River flood in 2005 shows (Tennberg et al. 2010).

Local adaptations take place within the broader context of past and current political, economic, and social developments and decisions. "It is increasingly not enough", claims Newell, "to study actors and institutions that identify themselves as environmental in isolation from the global economic processes. These actors, institutions and practices are embedded in global economic processes" (Newell 2008). The political economic approach to adaptation examines political, economic, and societal contexts for adaptation at multiple levels of action. In the particular case of flood protection management, the extreme events and their societal significance reflect national, regional, and international practices of neoliberalism (Fieldman 2011). Neoliberalism re-organizes relationships and responsibilities between states and societal actors and aims at aligning governmental and local interests through reconfiguration of local identity and agency. It emphasizes local responsibility and innovation (Felli and Castree 2012; Granberg and Glover 2014; Fieldman 2009, 2011).

For flood protection management, this means renegotiating responsibilities between various actors involved with flood protection, emphasizing local responsibility and social innovations, to tackle flood risks. On the other hand, neoliberalism as a paradoxical phenomenon impairs coordinated collective action through a well-functioning state, which is considered critical to effective adaptation (Fieldman 2011; McLure and Baker 2013).

Flood discourses in Finnish Lapland are mostly technical by nature, with questions of improving flood warning and observation systems, developing risk assessments and maps, and identifying the most cost-effective flood prevention measures. Such discourses are powerful in developing understanding of what is the problem, why action is needed, and what type of action is possible to address the problem. However, this depoliticization of societal problems and governance makes it free from public scrutiny and participation, and free from responsibility. From the perspective of the political economy, such depoliticization is an effort to settle political questions in a "technical" manner. The political economy approach challenges this. By challenging the technical discourse on floods, that approach provides "us with an improved and less bounded sense of who governs and on whose behalf, how they govern and the implications of those practices of governing, in social and environmental terms" (Newell 2008).

The new legislation with emphasis on local responsibility for flood protection requires land and property owners, city authorities, companies, and regional authorities to discuss and renegotiate responsibility. Responsibility is a core issue in adaptation governance, i.e., "How is the responsibility to adapt and to implement adaptation measures shared between various levels and actors in adaptation governance?" Questions about what the problem is, how it is to be governed, and by whom, all entail responsibilities. Beck (1992) claimed that it is typical of environmental problems and their management in modern societies that responsibility for them "disappears" into structures of political governance. As a result, we have widely spread "organized irresponsibility," which can be attributed to the nature of the problems in such societies, for they are not easily defined, assessed or explained. Moreover, the solutions require actions by many. Often, as related to practices of organized irresponsibility, risks are downgraded, and easy, quick and cheap solutions are preferred over difficult, time-consuming, and expensive alternatives. The two cases of flood protection in Lapland allows us to trace the processes of defining, bearing, and sharing responsibility for flood protection management.

13.4 Kemi River Case

Kemi River is the longest in Finland, and flows through Rovaniemi. The most recent major flood took place in 1993, which had a typical water level, occurring once every 20 years. In the case of an extremely rare flood, i.e., once every 250 years, up to 10,000 people live in the flood risk area near the city center. In a once-every-100-year flood, 1115 people, 305 houses, and 815 buildings are threatened. In a once-every-250-year flood, 2600 individuals, 710 houses, and 1705 buildings are in danger. In addition, in the flood-prone area called Saarenkylä, there is a healthcare center, daycare centers, a home for the elderly, culturally important places, major local tourist attractions, and industrial activities, which during flooding could sustain environmental damage. In the once-every-century flood, the amount of damage is expected to be about 23 million Euros and, for a once-every-250-year flood, this would reach 76 million Euros (Lapin ELY-keskus 2015a).

The flood group proposes a once-every-250-year solution to flood management, i.e., flood-prone areas will be protected from very rare floods during that interval. The standard flood protection level is once-per-century, which protects the area from major floods once during that period. The appropriate level of protection from floods was discussed by the flood group. Greater flood protection is usually reserved for sites that are difficult to evacuate (hospitals, day cares, and homes for the elderly), but in Rovaniemi such protection has been extended to ordinary houses. The group also proposes as the primary measure for flood protection a new reservoir, called Kemihaara, to an EU NATURA 2000 protected area. This area is about 80 km from Rovaniemi, in the upper stream of the Kemi River, in an area belonging to another municipality, Pelkosenniemi (see Fig. 13.2). A secondary measure would

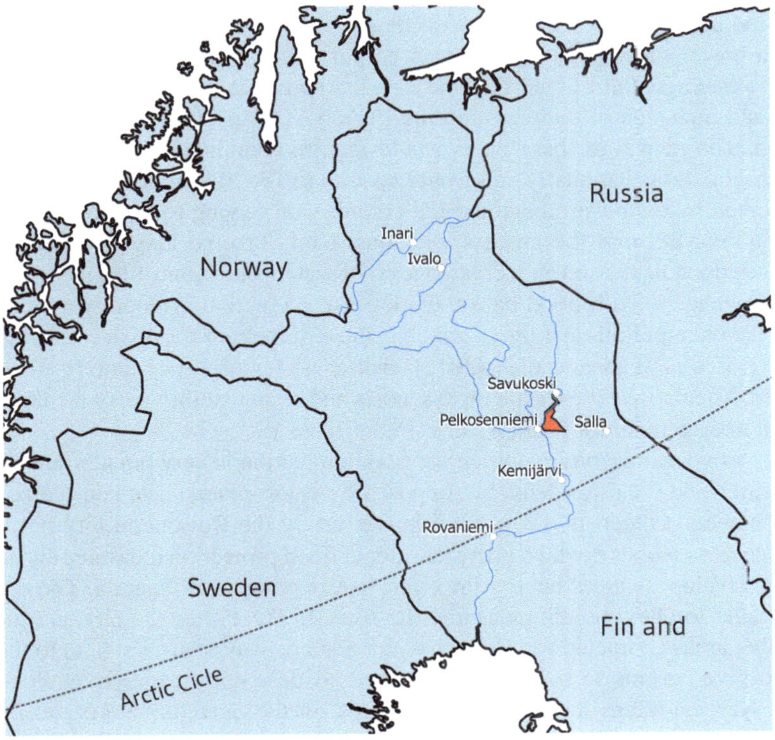

Fig. 13.2 Map of Kemihaara reservoir (marked in *red*)

be emergency elevation of the water level upstream in Kemi Lake (in the municipality of Kemijärvi), and the building of flood banks in flood-risk areas of Rovaniemi (Lapin ELY-keskus 2015a).

Because of the high target level of flood protection (once-every-250-year flood instead of once-every-century), many available flood protection measures, such as allowing a rise of water level in Kemi Lake in case of emergency, building new flood banks, river dredging, and temporary flood protection measures, are considered inadequate. Reasons for the high risk-level assessment are many. First, the location of the city at the junction of two major rivers (the Ounas and Kemi) poses a risk for major floods. Second, there is also the probability of material damage and costs related to floods in the region. Third, there are activities and locations that are considered difficult to evacuate in case of a flood (Ehdotus Kemijoen vesistöalueen tulvariskien hallintasuunnitelmaksi 2016–2021). This is a somewhat controversial issue because, at least for the moment, city planners in Rovaniemi still use the once-per-century level + 0.5 meter-level as a standard for planning new residential areas along the Kemi River, for example in the Vaarala area (Rovaniemen kaupunki 2012). One would expect that under these circumstances, the city would demand a higher level for planning new residential areas as the risk of rare floods has been stressed in the flood management planning.

The proposal prioritizes a reservoir as a solution for flood protection and supports the use of renewable energy in the region, because the reservoir could also have a power plant. This idea, known earlier as the Vuotos reservoir, originated 40–50 years ago, but has not become a reality owing to negative water court decisions and inclusion of the region in the EU's NATURA 2000 nature conservation system. The revival of these plans would require complicated administrative and legal negotiations nationally and would take up to 18–20 years to become a reality if accepted by the government and EU commission (owing to EU NATURA 2000 regulations). Because the process will most likely take so long, the flood group proposes flood banks and an increased level of water (maximum 1 m) in Kemi Lake, for emergencies. With flood banks, the level of protection in flood risk areas may reach the once-per-50-year flood, which is the requirement for private insurance. In the city area new general plan (2015), risk areas for once-per-century floods are marked. The overview over the means, costs and responsibility party for flood protection management for Kemi River is in the Table 13.1.

The proposed reservoir solution for flood protection is very popular among city authorities and local residents but opposed by environmentalists and the regional environmental center. It is a solution preferred by the Rovaniemi city because it saves costs as it does not take many additional flood protection measures and allows city to develop its areas nearby the river, like in recently in Vaarala. The city has been eager to allow buildings nearby the river. In the 1990s, the city and the surrounding areas competed for new taxpayers and one way to attract them to the area was to give permits to build houses in areas widely known locally as flood risk areas. New buildings in the flood prone area, in the Saarenkylä area, were built according to existing requirements and instructions of those days. Since 1999, the regulations have been tighter. The city, however, is not willing to cover expenses for

Table 13.1 Flood protection alternatives in Rovaniemi

	Impact on the protection level	Cost	Responsibility
Water reservoir	High impact, up to 1/250, harmful environmental effects	250 million euros	Hydro power company
Flood bank	Considerable impact, up to 1/100	13 million euros	Rovaniemi municipality, house and property owners
Emergency raise of water level in Kemi Lake	Considerable impact, up to 1/100	1–5 million	Municipality, Lapland trade, Traffic and Environment center
Dredging	Impact uncertain, harmful environmental effects	55–165 million euros	Lapland trade, Traffic and Environment center
Temporary structures	Not very effective	1 million euros	Municipality, house and property owners

Source: Lapin ELY-keskus (2015a)

extra measures needed to protect the flood risk areas with houses and other activities. According to the head of the city, the expenses are too high for the city to cover on its own, and no financial support is expected from the state. The city's view is that the focus of current discussion should be on finding efficient means for flood protection, not discussing responsibility for the problem.

Local residents have been involved in the flood protection planning through information and discussion events, email and the Internet, and surveys. Further, the local resident association in Saarenkylä was active in organizing their own events. For residents, especially in Saarenkylä and other flood-prone areas along the river, the new reservoir is a welcome solution. There is much skepticism regarding the efficiency of the flood banks, because of the soil structure. In addition, the flood protection measures are costly, require cooperation and, to be efficient, they must be extensive (e.g., building flood banks along the riverside). Building a new reservoir would mean less pressure on local inhabitants and their associations to become active in building those banks. It appears that knowledge about the new law, insurance, and its requirements has not been widespread among property owners in Rovaniemi. Further, it is somewhat uncertain what the companies actually offer as services and insurance products, as well as their prices and relationships to other insurance policies.

The central figure in the present flood case is the hydropower company. With its help, the responsibility to deal with the flood risk and finding its solution will be externalized in space, time, and agency. The climate argument is used to support this solution, i.e., more hydropower will be needed in the future owing to the change from non-renewable to renewable energy resources. If the plan materializes, the reservoir will be built in another municipality's area (Pelkosenniemi), with significant effects locally. Therefore, it is resisted by most municipalities affected by the reservoir in eastern Lapland, except Salla. The reservoir could be built in about 18–20 years if all national and EU legal and administrative constraints from negative water courts decisions are overcome. NATURA 2000 restrictions and nature protection measures in the area were originally targeted at Vuotos reservoir. The hydropower company prefers this solution, now called Kemihaara because the responsibility for the reservoir, its operation, and its benefits are assigned to the company. The widely held assumption among city authorities and local residents is that the company will be interested in developing the reservoir plan on its own and funding it with its own resources, even if implementation would occur in only 18–20 years. The company has declared that it has no "active" plans concerning the reservoir.

Privatization of flood risk and its management has been partially successful in regard to the introduction of greater local responsibility over flood protection and of market-based methods, insurance companies and their services, as a means for flood damage compensation in lieu of state-paid compensation. There is a re-division of risk and responsibility currently taking place in Rovaniemi because of the new legislation, which emphasizes individual and municipal responsibility. The risk is now mostly placed on local property owners and if the reservoir plan survives all the legal and administrative obstacles, the risk will be carried by the hydropower company. So far, the city appears to have been successful in avoiding responsibility, giving substantial responsibility for flood protection to the company to execute the expected measures. The mix of substantial expectation of local flood risk combined with an effort to manage that risk by externalizing it to the energy company is perplexing, and not even very logical or rational. However, this is attractive for the ambitions of the political elite in Lapland and is somewhat understandable. The old dream of Vuotos reservoir is now one step closer to being realized.

13.5 Ivalo River Case

The village of Ivalo, with 3600 inhabitants, is situated on both banks of the lower Ivalo River in the municipality of Inari. That river is known as "a bad flooding river" in Finnish Lapland, with recurring floods in spring. Major spring floods have occurred in 1966, 1968, 1981, 1993, 2000 and 2005. Most villages in the Inari area are along the river, Inari Lake, and smaller lakes. The area between Ivalo village and Törmänen is densely built. In the flood risk area there are 85 people, 55 houses, and 150 buildings threatened by a once-per-century flood, and 2105 people, 805 houses and 1535 buildings threatened by a once-per-250-year flood. The center of the village is protected by flood banks from once-per-century or more frequent floods. In addition, outside the center, there are flood banks with protection from once-per-50-year floods. If a once-per-century flood occurs, material damage (estimated to be ~3.2 million Euros) in Ivalo would be considerable, with road connections cut. There may also be a need to evacuate a healthcare center, service center, daycare facilities, and important service providers, because power stations, electricity facilities and waste water systems may be threatened (Lapin ELY-keskus 2015b). The overview of means, costs and responsible party for regional flood protection in Ivalo River is in the Table 13.2.

Table 13.2 Alternative means of flood protection, costs and responsibility

	Impact on the protection level	Cost	Responsibility
Sawing of river ice	Moderate impact	20,000–60,000 euros/year	Lapland economic development, Traffic and Environment center
Measures by house and property owners	Moderate impact	115,000–700,000 euros	House and property owners
Flood banks	Considerable impact	3,2–5 million euros	Municipality, Lapland economic development, Traffic and Environment center, house and property owners,
Canal	Low impact, not effective	12–15 million euros	Authorities
Flood pool	High impact, but difficult to execute, harmful environmental effects, expensive	15–30 million euros	Authorities
Dredging	Low impact, negative environmental impacts	25–30 million euros	Authorities
Roads	Important connections saved in a case of 1/250 flood	1,7–3,4 million euros	Inari municipality, Lapland economic development, Traffic and Environment center

Source: Lapin ELY-keskus (2015b)

The Ivalo flood group considered as a goal to protect Ivalo village from floods once per century, and various ways to help tackle the risk in the near future. Flood protection measures based on flood banks were started beginning in 1985, and were added to in 1993, 2000, 2003, and 2005. Currently, there are about 13 km of flood banks in the Ivalo village area. According to a management proposal, there is a need to make the banks higher (0.1–0.4 m) for about 4 km, to reach the once-per-century protection level everywhere and, in some places, such as near the power station, to the level of once-per-250 years. The flood group proposes that the current flood banks be measured and built higher as needed, via collaboration between the municipality of Inari and property owners. Also, the need for new banks will be assessed by the municipality. Another priority measure is to raise the level of the roads and strengthen them to endure once-per-250-year floods. This measure addresses 7 km of roads. This would be the responsibility of the Inari municipality and the Center for Economic Development, Traffic and Environment. Also, the sawing of ice and destroying of ice dams in the river, means used in current flood protection, are listed as secondary flood protection measures as a responsibility of the Center for Trade, Traffic and Environment (Lapin ELY-keskus 2015b).

With these measures, i.e., the elevation of existing flood banks and building new ones, almost all buildings and activities can be protected in the Ivalo flood risk area. The proposal acknowledges that flood banks are a locally controversial issue.

They are suspected to be too fragile to survive the water volumes during floods, and also directing water along hazardous courses in the village area. In public debates, other measures have been proposed and discussed, namely, temporary flood protection measures, dredging, a flood pool, and a canal. Also in the debates, the concept of letting floodwaters disperse along the lower river course appeared to be a relatively easy solution to implement in Ivalo, with its sparse population and large natural areas without build-up (Vuojala-Magga and Turunen 2013). The flood group proposal does not include such measures, including a canal, dredging, or flood pool. According to that group, these measures would have either temporary or permanent environmentally hazardous effects compared with flood banks, elevation of road level, and sawing of ice. The proposal by the local residents to let the river run freely or allow it to run along the old stream is considered to have such poor flood protection that it was not considered in the assessment of alternative flood protection measures. In general feedback about the plan, the regional environmental organization complimented the plan for not proposing a new flood pool or dredging because of their considerable environmental effects. Moreover, the reindeer herders' association stated that the plan is realistic and feasible in terms of the means chosen (Ivalojoen tulvaryhmä 2015).

13.6 Renegotiating Flood Risk and Responsibility to Protect

The new legislation on flood compensation in Finland, i.e., its privatization, has meant renegotiating responsibilities for flood protection between local actors, regional environmental centers, municipal and regional authorities, companies and other related actors such as rescue services. The new flood legislation has since 2010 allowed regions to independently negotiate the level of risk and preferred measures of flood protection, in the absence of clearly set national standards. This freedom is reflected in the development of the two flood management plans. They have very different solutions to the problem of defining risk level, criteria for assessing effective measures, and the sharing of responsibilities in flood protection. The main points of the finalized regional flood management plans can be found in Table 13.3.

In Ivalo, the risk level was set to secure the village from very rare floods occurring once every 100 years. In Rovaniemi, that level was set to floods once every 250 years, which are classified as extremely rare. The proposed solutions also follow different logic. In Rovaniemi, the preferred solution is a new reservoir to be built 80 km from the city, developed by the hydropower company. If the legal and administrative constraints are addressed, the earliest the reservoir would begin to construct is 18–20 years. Meanwhile, for major floods, the city will rely on flood banks and emergency elevation of the water level in Kemi Lake. In Ivalo, the solution is to extend and strengthen flood banks, and to rely on the responsibility of house and property owners and their collaboration with the municipality for flood preparation. In addition, roads will be improved to endure extremely rare floods and the sawing of river ice will continue.

Table 13.3 Overview of flood protection solutions in the two plans discussed

	Rovaniemi, Kemi River	Ivalo River
Time	In the far, uncertain future, in 18–20 years	In the near future
Space	Water reservoir and emergency water level rise in Kemi Lake, 80 km away, situated in other municipalities, and some local measures, by building flood banks	Inside the village, making existing flood banks higher, and building new flood banks; raising and strengthening of major roads; and continuing sawing of ice when needed
Level of risk	Once in 250 years	Once in 100 years, in some place once in 250 years
The cost	250 million euros	5–6 million euros
The responsible party	Hydro power company, other municipalities, regional authorities, local residents, and city authorities	House and property owners, local and regional authorities
Responsibilization	Avoiding local responsibility by resisting flood banks, and to externalize flood protection measures outside the city	Taking local responsibility

These two proposals for flood protection management highlight the logic of neoliberal adaptation to changing conditions. From the governance point of view, adaptation is a particular milieu made up of responsibilities, which is spread among various stakeholders involved with flood protection, including local residents, local, regional, and national authorities and other stakeholders. The flexible approach within the EU and national flood protection management, emphasizing attention to local conditions and lack of clear national guidance on criteria for target levels of protection, led to very different solutions in risk assessment and identification of responsibilities in the two cases studied herein.

The neoliberal logic of responsibilization works as a particular, economically driven technique of power by spreading risks among different societal actors and by privatization of responsibility to tackle those risks. Economic rationality worked in various ways in the two cases as a rule for sharing risks and responsibilities. For the Ivalo River, the new management plan relies on the economic rationality of house owners and municipalities to advance their responsibilization. The Ivalo flood management plan opts for a relatively easy, cheap and effective measure for flood protection. In Rovaniemi, the economic rationality works in another way. Local residents and city authorities are eager to avoid costs related to flood protection, and prefer to have the hydropower company take the leading role and responsibility in identifying and funding a means for effective flood protection in the region. The role of the participatory process and local political pressure also varied in the two planning processes. In Rovaniemi, residents, especially those living near the river, widely supported the proposed flood protection measures. In Ivalo, the popular idea of letting floodwater run freely and abolishing flood banks was not included in the planning process for further investigation. In the first case, the popular opinion supported the local authorities' preferences, and was clearly noted. In the second case, the popular opinion challenged local decision-makers' preferences and was eventually dismissed. For Ivalo River, the flood risk management plan is now accepted and in force, while for Kemi River, the plan is only partially accepted. The debate about the new reservoir continues as lively as before.

References

Agrawal A (2005) Environmentality: technologies of government and the making of subjects. Duke University Press, Durham/London

Beck U (1992) Risk society. Sage, London

EU Flood Directive (2007) http://eur-lex.europa.eu/legal-content/EN/TXT/PDF/?uri=CELEX:32007L0060&from=BG. Accessed 13 May 2016

EU Water Directive (2000) http://eur-lex.europa.eu/resource.html?uri=cellar:5c835afb-2ec6-4577-bdf8-756d3d694eeb.0004.02/DOC_1&format=PDF. Accessed 13 May 2016

Felli R, Castree N (2012) Neoliberalising adaptation to environmental change: foresight or foreclosure? Environ Plan A Int J Urban Reg Res 44(1):1–4

Fieldman G (2009) Climate change: global risks, challenges and decisions. IOP Conf Ser Earth Environ Sci 6. doi:10.1088/1755-1307/6/0/102006

Fieldman G (2011) Neoliberalism, the production of vulnerability and the hobbled state: systemic barriers to climate adaptation. Clim Dev 3(2):159–174

Finanssialan keskusliitto (2014) Vesistövahinkojen korvaaminen kotivakuuksista. https://www.fine.fi/media/julkaisut-2013/vesistotulvavahinkojen-korvaaminen-kotivakuutuksista-2014.pdf. Accessed 13 May 2016

Granberg M, Glover L (2014) Adaptation and maladaptation in Australian national climate change policy. J Environ Policy Plan 16(2):147–159

IPCC, International Panel on Climate change (2014) Annex II glossary. In: Climate change 2014: impacts, adaptation, and vulnerability. Part B: Regional aspects. Contribution of working group II to the fifth assessment report of the Intergovernmental Panel on Climate Change. http://www.ipcc.ch/pdf/assessment-report/ar5/wg2/WGIIAR5-AnnexII_FINAL.pdf. Accessed 13 May 2016

Ivalojoen tulvaryhmä (2015) Pöytäkirja 8. kokouksesta 11.6.2015. http://www.ymparisto.fi/download/noname/%7B95685D63-F521-425F-BCB2-0F58FECCF6E5%7D/110118. Accessed 13 May 2016

Keskitalo C, Juhola S, Westerhoff L (2011) Climate change as governmentality: technologies of government for adaptation in three European countries. J Environ Plan Manag 55(4):435–452

Lapin ELY-keskus (2015a) Ehdotus Kemijoen vesistöalueen tulvariskien hallintasuunnitelmaksi. http://www.ymparisto.fi/download/noname/%7B85995250-3104-47DA-BBF7-7A4193715BE7%7D/112141. Accessed 13 May 2016

Lapin ELY-keskus (2015b) Ehdotus Ivalojoen vesistöalueen tulvariskien hallintasuunnitelmaksi. http://www.ymparisto.fi/download/noname/%7B84EB529C-47EE-4A27-AAE8-C4C56C251F12%7D/112132. Accessed 13 May 2016

Law on flood risk management (2010) Laki tulvariskien hallinnasta. http://www.finlex.fi/fi/laki/alkup/2010/20100620. Accessed 13 May 2016

McClure L, Baker D (2013) Doing adaptation differently? Does neoliberalism influence adaptation planning in Queensland? In: Ruming K, Randolph B, Gurran N (eds) State of Australian cities conference 2013: refereed proceedings. State of Australian Cities Research Network, Shangri-La Hotel, Sydney

MMM, Ministry of agriculture and forestry (2015) Decision. http://mmm.fi/documents/1410837/1516651/Tulvariskien_hallintasuunnitelmien_hyv%C3%A4ksyminen_p%C3%A4%C3%A4t%C3%B6s.pdf/67102d0c-130c-4647-b762-2b0b79b8f28d. Accessed 13 May 2016

Newell P (2008) The political economy of global environmental governance. Rev Int Stud 34:507–529

Parjanne A, Huojuna M (2014) Tulviin varautuminen rakentamisessa. Ympäristöopas. https://helda.helsinki.fi/handle/10138/135189. Accessed 13 May 2016

Raitio K, Rannikko P (2006) Metsien käyttö ja sosiaalinen kestävyys: Metsähallituksen roolin muuttuminen Lieksassa. Metsätieteen aikakauskirja 2:271–292

Rannikko P (1997) From functional to symbolic local community. A case study of a forest village in eastern Finland. Res Community Sociol 7:223–246

Rovaniemen kaupunki (2012) https://www.rovaniemi.fi/loader.aspx?id=850c77a9-32ab-4e67-9e3c-922be04d61ce. Accessed 13 May 2016

Sinevaara-Niskanen H, Tennberg M (2012) Responsibilisation for adaptation. In: Tennberg M (ed) Governing the uncertain: adaptation and climate in Russia and Finland. Springer, Berlin, pp 125–136

Suurtulvatyöryhmä (Timonen R, Ruuska R, Suihkonen K, Taipale P, Ollila M, Kouvalainen S, Savea-Nukala T, Maunula M, Vähäsöyrinki E, Hanski M) (2003) *Suurtulvatyöryhmän loppuraportti.* Maa- ja metsätalousministeriö, Helsinki. Työryhmämuistio 2003:6 http://wwwb.mmm.fi/julkaisut/tyoryhmamuistiot/2003/tr2003_6.pdf. Accessed 13 May 2016

Tennberg M, Vola J (2014) Myrskyjä ei voi hallita: haavoittuvuuden poliittinen talous. Alue Ympäristö 43(1):73–84

Tennberg M, Vuojala-Magga T, Turunen M (2010) Ivalo River and its habitants: there have always been floods – what is different now? In: Hovelsrud G, Smit B (eds) Community adaptation and vulnerability in Arctic regions. Springer, Berlin, pp 221–237

Tulvariskityöryhmä (Kaatra K, Hanski M, Hurmeranta U, Madekivi O, Nyrcos H, Paunila J, Routti-Hietala N, Ruuska R, Salila J, Savea-Nukala T, Tynkynen A, Ylitalo J, Kemppainen P, Rotko P) (2009) *Tulvariskityöryhmän raportti.* Maa- ja metsätalousministeriö, Helsinki. Työryhmämuistio 2009:5. http://www.mmm.fi/fi/index/julkaisut/tyoryhmamuistiot.html. Accessed 13 May 2016

Tulvavahinkotyöryhmä (Kaatra K, Suihkonen K, Tolvi T, Välipirtti KL, Leinonen A, Reskola V-P, Ollila M, Kujanpää M, Hurmeranta U, Gullstén N, Ijäs H, Seppänen R, Valjakka K) (2006) *Tulvavahinkotyöryhmän loppuraportti.* Maa- ja metsätalousministeriö, Helsinki. Työryhmämuistio 2006:16. http://www.mmm.fi/fi/index/julkaisut/tyoryhmamuistiot.html. Accessed 13 May 2016

Veijalainen N, Jakkila J, Nurmi T, Vehviläinen B, Marttunen M, Aaltonen J (2012) *Suomen vesivarat ja ilmastonmuutos – vaikutukset ja muutoksiin sopeutuminen.* Water Adapt-projektin loppuraportti. Suomen ympäristö 16/2012. https://helda.helsinki.fi/bitstream/handle/10138/38789/SY16_2012.pdf?sequence=1. Accessed 13 May 2016

Vuojala-Magga T, Turunen M (2013) Experiences with an Arctic river – flood prevention in the town of Ivalo. In: Keskitalo C (ed) Climate change and flood risk management: adaptation and extreme events at the local level. Edward Elgar, Cheltenhamn, pp 237–270

Summary: Social Adaptation to the Changes of the Terrestrial Ecosystem in Eastern Siberia, with an Emphasis on Water Environments

A brief summary of this book follows.

The extent of Arctic summer sea ice has been decreasing, especially on the Eurasian continental side. Global warming is a partial cause. Cyclones have appeared frequently in summer in the region, bringing increased precipitation to Siberia in particular. Meteorological data reveal high rates of summer precipitation in the upper and middle parts of the Siberian Lena River Basin from 2005 to 2008, and in 2012.

Summer river flooding around Yakutsk, capital city of the Sakha Republic (Yakutia) of the Russian Federation, has become a problem, severely damaging local agriculture and rural communities. Ice-jam-related disasters also occur along the Lena River in spring, and can be severe when low winter temperatures are followed by gradually increasing spring temperatures. Such spring river floods have caused severe damage to the property of residents living along the river almost every year since 1998.

Local perceptions and local government adaptation strategies were investigated for both spring and summer river flooding. Spring river flooding has been recognized as beneficial except when it causes damage to villages along the river. This is because spring floods bring nutrient-rich water to the river islands on which farmers cultivate pasture for cattle and horses. Summer river flooding, on the other hand, is seen as a hazard because it submerges the pasture completely and prevents harvesting of hay for cattle and horses.

Village relocation is one adaptation strategy to prevent damage from spring floods. Because people prefer to live along the river, on which their subsistence depends, they have agreed to migrate seasonally with government support. However, there have been no similar adaptation strategies implemented in relation to summer flooding. Based on our observations and analysis, we intend to promote sustainable subsistence activities in the region. We will do so by proposing strategies to facilitate information transmission and to improve feed-hay distribution networks, which can aid in adaptation to spring and summer river flooding.

T. Hiyama, H. Takakura (eds.), *Global Warming and Human - Nature Dimension in Northern Eurasia*, Global Environmental Studies,
DOI 10.1007/978-981-10-4648-3

We also investigated how livestock farmers and hunters have adapted to social-environmental changes in the region. Interviews with keepers of domesticated reindeer revealed that current climate change has not severely impacted their operations. Careful management of the microhabitats used by domesticated reindeer has allowed them to successfully adapt to climate change even though they were severely affected by social changes following the collapse of the Soviet Union.

Although chapters describing wild reindeer behavior are not included in this book, we have documented the migration routes of wild reindeer, based on tracking them with an ARGOS satellite system, to understand their seasonal behavior. Similar to reindeer populations in North America and North Europe, Siberian reindeer have a summer breeding season, winter hibernating season, and migrate during other seasons. Because recent climate change has led to degradation of reindeer moss in winter, the birthrate and weight of reindeer in the spring has tended to decline. Establishment of protected winter hibernating grounds would help to protect wild reindeer populations. In order to preserve the practice of keeping reindeer, one of the primary subsistence activities in Siberia, we suggest that government subsidies should be provided to keepers of reindeers and to hunters of carnivores.

We would be grateful if our analyses and recommended adaptation strategies were disseminated to stakeholders in Siberia and considered by local governments and people in the region.

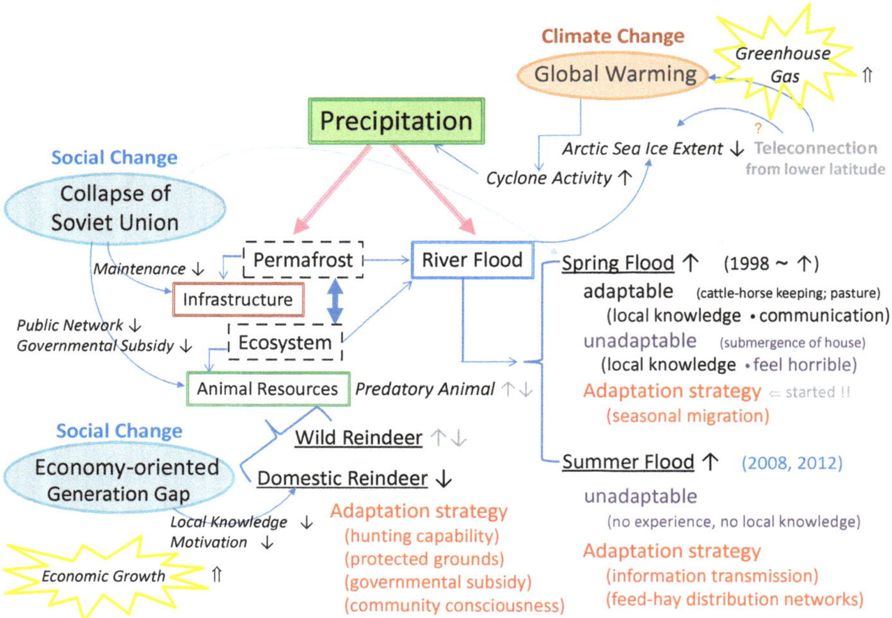

Interactions between humans and the environment in Siberia
The *left side* shows reindeer-related subsistence activities and the *right* indicates river flood impacts on local residents. In the *lower part* of the figure, several adaptation strategies are proposed in relation to these issues

MIX

Papier | Fördert gute Waldnutzung

FSC® C083411

Kolibri 360 GmbH
Ferdinand-Jühlke-Straße 7
99095 Erfurt, Deutschland
produktsicherheit@kolibri360.de